SHATTERING THE WALL

IMAGINE HEALTH CARE WITHOUT PREVENTABLE HARM

Written by: Anne Gunderson
and Tracy Granzyk
with David Mayer

ISBN: 978-1-4834-8450-1 (sc)
ISBN: 978-1-4834-8452-5 (hc)
ISBN: 978-1-4834-8451-8 (e)

Library of Congress Control Number: 2018904664

Because of the dynamic nature of the Internet, any web addresses or links contained in this book may have changed since publication and may no longer be valid. The views expressed in this work are solely those of the author and do not necessarily reflect the views of the publisher, and the publisher hereby disclaims any responsibility for them.

This book is a work of non-fiction. Unless otherwise noted, the author and the publisher make no explicit guarantees as to the accuracy of the information contained in this book and in some cases, names of people and places have been altered to protect their privacy.

Any people depicted in stock imagery provided by Getty Images are models, and such images are being used for illustrative purposes only. Certain stock imagery © Getty Images.

Lulu Publishing Services rev. date: 5/17/2018

CONTENTS

FOREWORD

The Academy for Emerging Leaders in Patient Safety: The Telluride Experience (TTE), formerly the Roundtable and Patient Safety Summer Camp, has inspired a new generation of medical and nursing students, residents, and other learners to understand why patient safety matters. They are introduced to essential concepts and practices to care for patients safely—every patient, every time.

TTE was conceived not long after the publication of the landmark Institute of Medicine (IOM) report, "To Err is Human," in 1999. For the first time in the history of medicine, physicians, nurses, and public members of the IOM panel acknowledged that although hospitals and other health care facilities do so much good every day to enable the sick to recover from illness and live longer, healthier lives, too many people suffer serious, preventable harm. At that time, no words existed to talk about patient harm. The traditional response when a patient was harmed or died was to move on to the next patient.

We have come so far since then. A new lexicon has been created: *medical error, failure-to-rescue, rapid-response systems, high reliability, just culture, safety culture, human factors,* and *root-cause analysis.*

These new words help us describe the phenomena of medical harm that all doctors and nurses observe in their training and professional careers. The words embody new knowledge that helps us understand the risks in the system and tools to mitigate and prevent them.

With new language has come a consciousness that compels us to fix broken systems of care. Many health care organizations track patient harm events, review them to learn why they occur, and implement improvements to prevent them from happening again.

Today, physicians, nurses, and other health care professionals are called upon to do more than provide competent care with compassion. They are expected to proactively identify unsafe situations and protect patients from falling into cracks in the system. Their job is to do the work—and improve it.

With the extraordinary leadership and dedication of Dr. David Mayer and Dr. Timothy McDonald, TTE convenes current and future leaders in patient safety, providing learners with the opportunity to engage with those in the forefront of making health care safe.

TTE is a forum where learners meet patients and family members who have experienced medical harm and have dedicated their lives to encouraging honesty in the aftermath of patient harm. Without honesty, success becomes a distant aspiration rather than a reality today.

As I wrote in *Wall of Silence: The Untold Story of the Medical Mistakes that Kill and Injure Millions of Americans*, we can't fix what we refuse to see, acknowledge, and own.

We have come so far, but we have many miles to go. At least the "wall of silence" has been shattered. The public knows that hospitals can be unsafe places. Nearly everyone has a story of medical care gone awry—or nearly so.

Because of The Telluride Experience, a cadre of physicians and nurses have their own stories of how they are making care safe. As the harm abates, let the stories, and the work, continue. There is no more noble cause in modern-day medicine.

—Rosemary Gibson

PREFACE

Anne Gunderson, Tracy Granzyk, and David Mayer

Welcome to *Shattering the Wall*. The book you are about to read is a collection of works created by young health care learners and the faculty members who taught them. Each chapter is deeply rooted in the health care domains of safety, quality, and leadership. The structure of the book is unique, as the foundational knowledge of the text came from our faculty, yet each chapter was inspired by our Telluride Scholars' reflections captured in the Telluride blog. The blog was created with the intention of providing an outlet where our learners could put their reflections from The Telluride Experience and the impact of its curriculum into their own words.

No matter what type of reader you may be health care professional, health science educator, or a patient who uses health services—this book will hopefully open your eyes to many of the intricacies of our current health care system, as well as the challenges faced by all of us in delivering safe, high-quality care. In this book, you will be exposed to firsthand accounts of real-life situations involving learners, professionals, patients, and their families, which reflect the historical background, current state, and lessons learned of each prevalent topic in health care practice.

Through The Telluride Experience, health care sciences learners attend a four-day immersion in patient safety and quality improvement. The purpose of the program is to fill gaps in health sciences education where formal, systematic patient safety and quality curriculum is lacking. The program identifies health science learners who are committed to developing patient safety and quality improvement of knowledge, skills, and behaviors that will prepare them to become future leaders who will make sustainable improvements in patient care throughout their careers. The curriculum is designed to ensure the educational content is relevant to all health science disciplines.

The future is limitless for these young leaders. The leaders of tomorrow have the opportunity to create a culture that is safe and, most importantly, patient-focused. By stretching the boundaries of traditional thinking and acting, learners and professionals can deliver higher-quality care to patients and their families, thus transforming and improving health care culture and practices for future generations.

We hope you will accept this work in the spirit in which it was intended: to give a larger platform to the voices of our future health care leaders, voices that will carry the much-needed health care culture change forward. Our faculty strongly believes in the "educate the young" premise—providing the future generation of health care leaders the knowledge, tools, techniques, and behaviors associated with high-quality, safe care.

The chapters do not have to be read in order. Feel free to pick a chapter on a topic that intrigues you—or start from the beginning and immerse yourself in a learning experience similar to what our learners experience while attending The Telluride Experience. Whether you are a health care learner, a health care professional, or a patient or family member interested in learning more about our current health care culture, you will appreciate *Shattering* because we all have one thing in common—we are all affected by the health of our current health care system, and we will all, one day, be patients.

Please note: Any and all proceeds from book sales will go directly to support future Telluride Scholars in their quest to become the safest, most patient-centered caregivers they can be. We are, in turn, grateful for your support.

CHAPTER 1

Not Alone

David Mayer

Blog Posts Written by Health-Care Learners

I Am Not Alone
Nicholas Clark
June 10, 2014

I have only just begun my journey here at Telluride, but one thing is clear: I am not alone. As the first day unfolded, I was given the opportunity to meet an amazing cast of future leaders, all of whom have patient safety as a main priority. Some of us are on the front end of our patient-safety journeys, while others are more accomplished. Some are beginning residency, while others are soon to graduate and take their Telluride experiences with them into their future careers. As the first and sole participant from my residency program, where our patient-safety revolution is still in its infancy, I take great pride and pleasure in knowing that I am not alone. Each one of us here at Telluride Patient Safety Resident Summer Camp is not alone. When we leave Telluride and head back to our home institutions, we will have a network of peers to call upon for advice and guidance on how to continue to make our patients and hospitals safer and free of unnecessary harm.

It's the System
Giovanna Sobrinho
July 27, 2017

I needed a few days after being at Telluride to let the experience settle in. It was a whirlwind of heart-wrenching stories, connections, and thought-provoking discussions. As a second-year resident, I came to Telluride with the baggage of a traditionally difficult intern year, the trauma of having lost my grandfather due to a multitude of systematic medical errors, and the viewpoint of having been a nurse prior to medical school. I needed to see a room full of people who didn't accept that "that's just the system we're in." I had grown tired and halfway accepted that answer, so it's amazing to see the influence that a group of inspired professionals can have. Every time I spoke to someone new, I was blown away by his or her dedication and drive. All the people in that room, from many types of training and different parts of the country, cared deeply about improving delivery of care for patients and taking care of each other as caregivers. There were not enough hours in the day for me to sit down and discover the unique background, impressive work, and innovative ideas that each person had to share. Empathy seems to be the common denominator that brought this group together. They were all deeply caring people who could not help but suffer when they came across the failures of our health systems. Highly empathetic people (as are most people in health care) need this type of nurturing environment to keep that part of themselves alive. I've been reenergized and will reach out to find similar people at my own hospital and see how far we can go.

Reflections on a New Generation of Caregivers

"I am not alone."

If there is one theme we have heard consistently from the young scholars who attend our TTE sessions through the years, it is that they now realize they are not alone—that there are others who also care deeply about preventable medical harm and the quality and safety we provide to the communities we serve. In the safe environment our sessions provide, they share their frustrations regarding caregivers who are focused on driving volume of care rather than quality of care, who are role-model physician-centric or nurse-centric rather than patient-centric principles, and who use intimidation and bullying when necessary to get their own ways. They share their fears of becoming that type of physician or nurse because the current culture supports that behavior. Each year, it is common to hear students and residents admit they are planning on leaving patient-care practice because they are so beaten down and disillusioned by what they have seen and experienced. They are looking for revalidation of the principles that made them want to be healers. They want to know they are not alone in having these feelings.

This disruptive culture in medicine has been well-documented in the literature. In a white paper titled *Unmet Needs: Teaching Physicians about Safe Patient Care* produced by the Lucian Leape Institute (LLI) and the National Patient Safety Foundation and published in 2009, the expert panel concluded this:

> All medical school deans and hospital CEOs need to declare and enforce a zero-tolerance policy for disrespectful and abusive behavior.
>
> The current hierarchical system of health care often tolerates disrespectful and abusive behavior toward students, residents, and others. Disruptive and abusive behavior takes many forms—as subtle as making a student feel foolish for asking a question or as overt as throwing instruments in an operating room.
>
> These behaviors create a culture of fear and intimidation, diminish pride and morale, impede learning, and sap joy and meaning from work. Disruptive and abusive behaviors are contagious. Abused students and workers perpetuate dysfunctional cultures by further modeling these behaviors. Changing this culture of fear and intimidation to a culture that facilitates and reinforces learning and respect is the responsibility of organization leaders.

We must change the current culture in health care if we are going to make safe, high-quality care a reality. But how do we raise the urgency and accomplish this culture change? It starts by "educating the young," a mantra used to describe how educators can empower health care learners to embrace a just culture and person-centered mentality, thereby equipping them with the assets to become leaders who will improve patient safety and the quality of health care.

The Institute of Medicine's "To Err is Human" report in 1999 estimated that between forty-four thousand and ninety-eight thousand patients die each year from preventable medical harm. Numerous follow-up studies found that, on average, one in ten patients admitted to hospitals suffers a serious, preventable adverse event at the hands of health care professionals. A study published in the *Journal of Patient Safety* by J. T. James in 2013 estimated deaths due to medical error were now thought to be four times the number revealed in the IOM report. And most recently, in 2016, an article by Makary and Daniel published in the *British Medical Journal* claimed that preventable medical harm was the third-leading cause of death in the United States.

Growing research and data also show that patients and families are not the only casualties of our current health care system. Caregivers also feel the effects of a broken system that needs a major overhaul. Depression, a loss of empathy, career burnout, and suicide are far too prevalent today in our health care workforce—our friends and colleagues who come to work each day trying to heal those we care for and do good.

Shortly after the IOM report was released, modification of health-sciences education dominated discussion in both the public and private sectors. Systematic safety and quality education for health care professionals was lacking, and the available literature provided no insight into how this could be done. Dr. Jordan Cohen, then president of the Association of American Medical Colleges, stated there needed to be a "collaborative effort to ensure that the next generation of physicians is adequately prepared to recognize the sources of error in medical practice, to acknowledge their own vulnerability to error, and to engage fully in the process of continuous quality improvement."

This urgent call for educational change was restated in the previously mentioned LLI white paper on medical education in 2009:

Health care delivery continues to be unsafe despite major patient safety improvement efforts over the past decade. The Roundtable concluded that substantive improvements in patient safety will be difficult to achieve without major medical education reform. Medical schools must not only assure that future physicians have the requisite knowledge, skills, behaviors, and attitudes to practice competently, but also are prepared to play active roles in identifying and resolving patient safety problems. While all health care professionals share responsibility for patient care improvements, the Roundtable recognizes that the medical profession plays a central role in encouraging and driving changes across all health professional education.

Unfortunately, seventeen years after the IOM report was released, calls for educational reform have gone unheard, and change has been absent. Despite continuing reports of overwhelming numbers of preventable patient deaths, a formal systematic patient safety and quality-care curriculum remains missing within health care educational programs across the country. Reforming medical education to adequately address safety and quality issues presents a major challenge to medical educators since the shortcomings that must be addressed are deeply entrenched in the tradition and culture of the institutions and organizations that compose the medical education system. Because of this old-world culture, medical education remains stuck in the 1980s.

As debate, ownership, and posturing continue to impede progress in our major medical and nursing teaching institutions, innovative health care educators are finding ways to educate and empower learners through novel patient-safety education opportunities. The Telluride Experience is one such entity, using a novel narrative and simulation-based curriculum to challenge young health care professionals to not only think differently about but also feel what it means to care for patients in a safe, high-quality health care environment.

The evolution of TTE began in 2004. I identified and invited select patient-safety and educational leaders to an inaugural patient-safety educational roundtable held in Telluride, Colorado. I chose that location because it would take faculty away from the daily pressures of work and place them in a relaxing, awe-inspiring, natural, think-tank surrounding that I hoped would unleash creativity and productivity.

But the key to TTE's success through the years was including the voices of patients—patients and family advocates who had experienced medical harm and personal loss—from the start. Over five long days and nights, attendees engaged in open and honest conversation and consensus building. Whether in the classroom, over a glass of wine during dinner, or out hiking in the morning before breakfast, the conversations during the week led to a well-thought-out patient-safety curricular framework. Each participant eagerly embraced the challenge and agreed to meet again the following year to continue this important work. The Telluride Experience Patient Safety Educational Roundtable had officially begun.

After our second year in Telluride, the multidisciplinary group had created a consensus patient safety curriculum that included eleven specific elements essential for effective patient safety education. Each member agreed to take the curriculum back to their home institutions and hopefully use it to create a new order in medical school and graduate medical education settings. Although some small gains were made, it still wasn't enough as most medical school and graduate medical education curricular maps were staying resistant to change.

In 2008, our Telluride faculty decided it was time to test their thinking with learners. Two medical students and two medical residents with strong interest in safety and quality were invited to join the roundtable. The young learners actively engaged in the discussions and, at the same time, learned from the experts. Their connection with our patient advocates during the week was especially rewarding as the students and residents quickly appreciated the importance of the patient voice in all our quality and safety work.

The success of this initial proof of concept model provided inspiration for the creation of a new Telluride Patient Safety program or training "Academy" for health science learners in 2009. To make the plan a reality, The Telluride Experience leadership sought funding through a small conference grant awarded by the Association for Healthcare Research and Quality (AHRQ). With this AHRQ financial support, the new Telluride

Experience Patient Safety program was officially launched with twenty interprofessional students.

Seeing similar successes with the twenty young health care leaders, in 2010 The Doctors Company Foundation (TDCF), COPIC, and the Committee of Interns and Residents all agreed to provide full scholarships to the Telluride Patient program for sixty medical students, nursing students, and resident physicians. The Telluride Experience vision and AELPS mission was now very clear—to offer an exceptional training immersion for young medical and nursing learners committed to becoming future leaders in reducing preventable medical harm and improving the quality of care provided to the communities we serve.

The TTE program continues to grow offering patient safety summer camps in three United States locations each year: Telluride, Colorado; Washington, DC; and Napa, California. Additional Telluride Experience programs are now also offered internationally for learners and future health care leaders in New South Wales, Australia, and Doha, Qatar. About two hundred future health care leaders attend one of our Telluride Experience Patient Safety programs across the country each year. Finally, in June 2016, all new incoming interns and fellows at MedStar Health—450 of them—went through the TTE program before they began working across the MedStar Health system on July 1.

What makes The Telluride Experience patient safety curriculum so unique and engaging is that the education shared with the young learners is built around four core themes:

- stories
- games and simulations
- patient and family partnership
- open and honest communication

Stories and narratives provide a true foundation for learning. Young learners through the years have had the opportunity to work closely with patients and families who have experienced medical harm. The Telluride Experience provided an opportunity to discuss medical errors with real patients and families so they could better understand the multiple breakdowns in care that led to that preventable harm. During these emotional learning sessions, their fears and concerns were brought to the forefront for open and honest discussion. Having patient advocates playing an important role in the learning process enhanced the discussions and ultimately transformed the Telluride learning experience into something very powerful and memorable for the young learners. Many developed long-standing relationships with our patient advocates.

Besides those four foundational themes, the curriculum includes in-depth training in the following domains:

- error science
- interdisciplinary teamwork and peer-peer communication skills
- effective communication skills related to unanticipated patient care outcomes, transparency, disclosure, apology, and early resolution models
- leadership, mindfulness, and professionalism
- informed consent/shared decision-making
- resilience science
- system error and human factors engineering and event-review methodologies
- creating a just culture within the framework of personal accountability
- "care for the caregiver" programs
- quality improvement techniques

Armed with this new knowledge and reenergized by the fact they are not alone and have a network of passionate colleagues like themselves, these young learners go back into their home environments and "infect" other young learners on the importance of patient safety, risk reduction, and quality improvement strategies for better patient care outcomes.

Taking The Telluride Experience Viral: The Telluride Blog

The reason we teach through the vehicle of stories is to harness the natural ability of narrative to carry a message that data, facts, and presentation slides often cannot. There has been a movement in health care toward the honoring of stories within the delivery of care for many reasons.

For example, health care professionals are finding that stories can help introduce difficult topics and bridge gaps between patient and provider experience and knowledge. Stories of patient and provider harm, like those shared at a Telluride Experience session, allow readers and viewers to live vicariously through the actions and experience of others in the hope that they will never make the same mistakes or be subject to a health care system that does not honor the spirit of a caregiver as well as their skills. Stories can also be shared within health systems in the form of good catches, near misses, unsafe conditions, or harm events as a tool for culture change, while at the same time, allowing for learning around the strengths and weaknesses lying deep within a health system before a patient or caregiver is harmed.

As we teach through stories, we also teach learners to value their own stories and the importance of reflective practice—that every experience they have in caring for others is both a valuable learning tool and a gift. Stories are embedded throughout the learning at TTE because they are naturally embedded in the profession. Using these stories as the foundation for TTE curriculum has time and again proven to make the learning stay with these young leaders in ways traditional lectures and presentations will never accomplish.

Since 2010, TTE scholars have been required to reflect on each day's learning through discussion of the day's events with colleagues and faculty, culminating in a written version in the form of a blog post on a topic of their choosing and whatever most left an impact on each of them. Further learning then occurs through the written reflection and the TTE blog, our foundation for *Shattering the Wall*, which now contains hundreds of posts by students and faculty reflecting on their time at a Telluride Experience Patient Safety program. Our TTE alumni stories and reflections were far too good not to share, and we knew a larger written work was needed to both aggregate and share their experiences with others. In the blog posts, many students openly share having witnessed patient harm at their hospitals, but they did not know who they could safely talk to or share their feelings about the event with, leaving them feeling despair and often depressed. Others have shared personal family loss due to preventable medical harm. Many blog posts are highly emotional and tear-jerking to read.

Over the course of the last seven years, the TTE blog has grown into a piece of the curriculum itself. Students are often writing their posts late into the evening after socializing with peers and carrying the day's learning and discussions to dinner in the old mining town of Telluride, the streets of Napa, nestled in the heart of wine country, and the bustling town that is downtown DC. Especially moving stories of courage are recognized at the start of each day to further embed the learning and support open and honest sharing in the classroom environment.

The TTE blog was created in 2010. We encourage those reading to go to The Telluride Experience website and peruse the blog posts written by our alumni; the content continues to build, and 2016 proved to be our largest number of attendees to date.

These young health care leaders learn they are not alone. They leave our patient safety summer camps reenergized and recommitted to their personal mission of remaining patient-centered and learning all they can about high-quality, low-risk patient care models. We have no doubt they will make a difference.

CHAPTER 2

The Heart Part: Using Stories to Change Minds
Tracy Granzyk

Blog Posts Written by Health Care Learners

Renewed Faith
Julie Morrison
July 11, 2012

There is nothing more powerful than a personal experience. Helen Haskell represented this at The Telluride Experience. Her son, Lewis, died of medical misdiagnosis that inhibited the team from providing appropriate life-saving treatment. His story was recounted in a video produced by Drs. Mayer and McDonald and the team at SolidLine Media. The story and the courage Helen has to continually attend the conference and provide a firsthand experience of an unsafe medical system was by far the most memorable thing about the week. There were lots of memorable things about the week, including the views from the gondola required to scale the mountains and the dedication of the faculty members, but Helen's story will continue to stay with me. I am hoping that her son's memory can act as a guiding force for all of us. A memory we can return to when we have lost our way along the road toward patient safety. A memory that will guide us back toward putting the patient and family as the center of our care if we go astray due to personal, family, financial, or other factors.

In the beginning of medical school, I was astounded by how many of the students were truly good people; they were caring people with big hearts and understanding minds. Research shows that our empathy decreases as our medical education progresses, and it reaches a nadir by year four. I can only assume that it continues to fall as we take on the responsibilities of internship and balance this with the desire and responsibility of starting a family, being a good neighbor, colleague, and son or daughter. I hope that Helen and Lewis's story counteracts this passivity and continues to call us toward our original goal of helping others, which, as the Hippocratic oath correctly details, first requires us to do no harm.

On neurology rotation, I saw a case that exemplified a large number of the topics we discussed at The Telluride Experience this year. A seventy-one-year-old woman that came in for an outpatient endoscopy with dilation for a postsurgical esophageal stricture. During the procedure, she desaturated and was intubated within minutes. The procedure was suspended, and a physical exam at the time showed no apparent life-threatening abnormalities. Within two hours, the physical exam showed anisocoria with the right pupil dilated to four millimeters. A stat CT scan of the brain showed "bubbles" within the cerebral veins as well as extensive cerebral edema and evidence of an uncal herniation with compression of the left cerebral peduncle. Mannitol was run, the head of the bed was elevated to thirty degrees, a central line was placed, the ventilator settings were changed to hyperventilation protocol and an FiO2 of 100 percent, and an extra ventricular drain was placed. Despite these measures, the patient continued to develop cerebral edema and had intracranial pressures into the triple digits throughout the night. She was declared dead by brain criteria the following day by our team.

I went home and cried that night. I was shocked by the thought that someone can come into the hospital thinking they are getting a benign routine procedure and die from massive cerebral edema within hours. The patient's husband was in the room the entire time; he was so strong to stand beside her during the interventions and stronger still for his open attitude toward the hospital staff. Luckily, he was not angry or resentful of the staff, and he believed that they did everything in their power to save his wife. This is lucky for him but also for the hospital because I don't think our staff is equipped to manage such events. I was waiting for the "safety officers" to come debrief the husband and explain the situation to him. I was counting the fifteen-minute deadline that they had and wondering how their involvement would have changed the course of events. I realized that these professionals could have a dramatic calming effect on the entire care team. During the intensive bedside interventions, the team was perplexed by the etiology of the patient's neurological deterioration. Many nurses were asking, "What happened?," "What's our story?," "What should I tell the patient's husband?" No one took the leadership role to answer these questions. In truth, no one knew the answer, but no one stepped up to simply say, "We don't know, but let's focus on stabilizing the patient and then we will investigate the cause."

I call this a poor outcome rather than an error because our attending found a report of twenty such cases in which air embolic strokes occurred during or after procedures that require insufflations (endoscopy, laparoscopy, colonoscopy, etc.). This then becomes a rare but known complication of the procedure, which made me think of the informed consent conversation we had during the session. I also thought of Michael Skolnik. I was fairly sure that no one informed the patient of the risk of air emboli infracting her brain. The pathogenic theory is that microscopic tears in the viscera allowed air to enter the esophageal veins, traveling to the IVC and through the lung vasculature to the systemic circulation (via pulmonary arterial-venous shunts).

We believe that the patient's desaturation was secondary to pulmonary emboli from the same mechanism. Does this represent a reasonable risk that the patient deserved to be notified of? I don't know. Twenty cases is not a large number, but the published case reports most likely underestimate the true incidence. Would I have wanted to be informed of this prior to the procedure? I don't know. I would have had to weigh the risk against the potential benefit. This is the essence of shared decision-making: giving the patient the facts and allowing them to ascribe their value assessment onto those facts and arrive at a logical decision. Without knowing this patient's baseline, I was left feeling only sad for the outcome and the loss of innocence that occurs when something so precious is lost—a life and a trust in the medical community.

I am glad that the patient's husband did not see it this way. I am glad that he maintained respect for, and trust in, the doctors caring for his wife. When my grandmother died of hepatic failure after taking an antibiotic, and the drug was subsequently pulled from the market due to risk of liver toxicity, I started to question my faith in the medical community. I became committed to the application of sound evidence to the relief of human suffering.

My faith in the medical community was restored slowly over my first two years in medical school as I learned that most physicians operate with very good intentions but in a complex environment that is not readily transparent or controllable. The derogatory-based jokes I heard during my third year, which I know were not ill-intended ones, started to erode that faith again. Attending TTE completely renewed my faith in the medical community. I was inspired by the fact that such successful professionals would take the time to teach medical students basic leadership and patient safety goals. In every conversation and action, I could tell that they had the patient's best interests as their primary goal. They renewed my faith in the medical community not only because they are true role models— individuals who I aspire to emulate—but also because they taught us how to lead the medical community into a more respectful, more patient-driven culture, and I now know we can do it!

The Heart Part of Transparency
Carey Candrian
June 17, 2012

On my trip home from TTE, I kept asking myself what would have been different if the session on respect and humiliation was covered earlier in the conference. These two concepts and processes are so central to the work we do and the work we want to achieve.

Throughout the week, we saw, heard, and felt what stories can do for the way we think, act, and make decisions. But telling stories demands trust, and it also demands humiliation because it exposes our natural limits as human beings, which can be incredibly uncomfortable. However, these moments of discomfort are often some of our most powerful learning tools because we open ourselves up temporarily to the possibility of change and transformation, whether we are the storyteller or the listener.

During the negotiation session, for instance, what would have been different if we underscored the importance of respect and humiliation during any and all negotiations?

What is it like to be in a negotiation? What does it look like? What does it sound like? And what does it feel like to have to let go of some of your own values and goals for the benefit of the collective, in this case, the patient and family? I think it would have been interesting to role-play during the negotiation session to take this concept from theory to practice in order to "try on" some of these skills and feel and talk about how difficult and uncomfortable this process can be—and how important and empowering it can be.

This move requires a lot of respect and a lot of humiliation in recognizing the heart part of this type of work, and even the heart part of talking about it with colleagues in the same room. But in doing so, it allows communication to be productive, meaning it allows us to achieve something new and different *together* that we normally couldn't have done on our own. And this begins to shift away from a more reproductive model of communication that essentially reproduces old and outdated skills, patterns, and habits that simply aren't conducive to our current health care situation.

Last week was without a doubt the most stimulating and enlightening learning experience I have ever had, so many thanks to all!

Faculty Reflections

This chapter will honor as many of The Telluride Experience Scholars' blog posts as possible, and as such, excerpts from their posts will be embedded wherever possible to emphasize the value of reflection and using stories to influence new ways of thinking throughout The Telluride Experience sessions. We begin with a reflection from Christopher Worsham, TTE Scholar, 2015.

> Stories can be great agents of change. As an internist, my day is spent hearing and telling stories—presenting a patient to an attending or a consultant is telling a story. Making a diagnosis is telling a story whose background, characters, and plot center around the central theme of a certain diagnosis. We are comfortable with these stories, we expect them to be told in a certain way, and we usually don't expect too many surprises … In day-to-day medicine, we often remove ourselves from the emotion of the events and stories unfolding in front of us. But when it comes to stories of patient safety, that emotion is unavoidable since we as providers are central characters in that patient story …

Stories of patients and caregivers alike have been the scaffolding that supports the curriculum at TTE around the world. Every TTE alumnus has watched the educational documentary *Tears to Transparency* film series, which includes the Lewis Blackman and Michael Skolnik stories. Every student is instructed to reflect on each day's learning, especially the stories told, and then write a story of their own in the blog post created solely for this purpose.

It's clear Worsham came to TTE with an innate appreciation of how the right narrative can influence the direction of every health care encounter for better or worse. The challenge at this early stage of professional life is to have the confidence, resourcefulness, and experience to draw on the stories all health care professionals have at their fingertips, and then to share them forward, in the right moments.

The perplexing dichotomy that Worsham hints at is that patients come into the health care environment with stories that are inherently loaded with emotion, and health care professionals are traditionally trained to stay objective, at a safe distance from emotions that could jeopardize the established hierarchy, revealing physicians and nurses as human. But times are changing, and stories making their way into the mainstream of health care could be partially responsible. Stories have the power to slide beneath the surface of our defenses undetected, challenging the comfortable status quo of our subconscious. As a result, the foundation of what the reader has long held as truth can be tested or even shaken to the core. Less of a dramatic, but nonetheless important, shift of consciousness occurs when health care professionals are simply reminded that they are human, overcoming training that often fails to honor the humanity at the heart of every health care professional. Padma Ravi, Telluride Experience Scholar, 2015, writes:

> Storytelling is a great way to learn, connect, empathize, and remember an incident for a lifetime. When we look at just statistics or plain charts and data, even though they draw your attention with regards to severity of a given situation they however fail to leave a lasting impression ... when someone tells you a story, they draw you a picture with words, they convey the strong emotions they have undergone and draw you into their lives by sharing something very personal and private. They not only connect to your heart and gut, but also leave a mark on your mind.

The goal of our TTE curriculum is to use stories that convey patient safety learning that connects the heart and the head, embedding knowledge that is both felt and absorbed on an intellectual level. Ravi is one of many TTE alumni who take away an understanding of the value in using this approach to educate medical professionals, and she articulates that intent almost verbatim in her blog post.

Stories by their very design help viewers or readers develop empathy, a much-needed quality for medical professionals who care for patients when they are most vulnerable. Research has shown that many medical students may experience a drop in empathy during their third year of medical school. This is also the time students leave the protection of the classroom and enter the clinics for rotations in different specialties, such as OB/GYN, psychiatry, surgery, and internal medicine. Many students interact with patients for the first time during this phase of training. If students aren't prepared for the storyline that is the actual delivery of health care versus the science of health care, the experience can be overwhelming.

Without a nurturing mentor to set expectations and manage the contradictions realized when seeing their career as imagined transform into the career as reality, students are often left to navigate these uncharted waters without a lifeboat. The business of health care delivery, more specifically academic medicine, includes an often-unfriendly hierarchy, a vicious pecking order, and a new low wo/man on the totem pole status that can lead to disappointment, depression, and even suicide in the very worse of scenarios. Giving young health care learners and professionals the ability to share their stories through The Telluride Experience blog lessens the burden of feeling alone and isolated and also provides a forum to share ideas, new ways of thinking, and collective reflections on the learning experienced. Stories have power to heal, help connect providers with patients and families, and offer that same healing and connection for and among medical colleagues.

Historical Background

The power of stories to drive change in health care has long caught on across the industry, and it has certainly been picked up by TTE learners as evidenced in their blog posts. Whether attendees come into the sessions with an appreciation for story-based learning or leave with having experienced it firsthand through the stories shared at TTE, their reflections certainly attest to the fact that these stories are contagious. Ravi writes:

> We kicked off our event with two powerful stories told by the surviving mothers of two patients who were victims of sentinel events. These stories were excellent learning opportunities. They were scary, but yet inspiring to do a better job at providing better quality of care. They have highlighted key aspects like the importance of timely communication, courage to speak up, and the pitfalls of working in silos and not including patients and their family members in the plan of care ...

The students' reflections provide anecdotal validation to a growing body of research supporting the how and why stories and narratives can be catalysts for behavior change. Evidence defining the science behind stories has also taken hold in businesses big and small, with marketing dream teams asking, "What's your story?" Venture capital firms dole out big money to start-up organizations that are successful in selling the story of their entrepreneurial dreams.

Health care, while slow to take on new business practices as a rule, has caught on to the transformative power of stories and has begun to capture and use the stories originating within their systems or practices. In the business of caring for people, stories abound, and these health care stories have the potential to connect patients and their care teams more deeply than lab reports, research, and abstract care plans. These stories also serve as reminders to health care professionals about what is really important in the delivery of care. They reset our collective moral compass, defining once again a true north for a profession that struggles to keep balance as its very landscape shifts.

According to Jonathan Gottschall, author of *The Storytelling Animal*, "Stories have an uncanny ability to mold our thinking and behavior." Fiction often proves to be more powerful in changing attitudes and beliefs than persuasive arguments. He says:

> Until recently we've only been able to speculate about story's persuasive effects. But over the last several decades, psychology has begun a serious study of how story affects the human mind. Results repeatedly show that our attitudes, fears, hopes, and values are strongly influenced by story. In fact, fiction seems to be more effective at changing beliefs than writing that is specifically designed to persuade through argument and evidence.

Good stories, built on classic narrative structure, can sneak past our defenses and long-held beliefs, introducing ideas that might be rejected if presented as fact with statistics as backing, a tactic traditionally employed in hospital boardrooms and departmental meetings.

Current research on the use of narrative in medicine has shown we can use stories to encourage the adoption of positive health behaviors in cancer-prevention strategies or even to improve blood pressure. Research also shows the stories we pass on to one another within our social networks possess the power to influence change in health-related behavior. Importantly, stories have been shown to create change on a physiological level, increasing neurotransmitter levels like dopamine, oxytocin, and cortisol. By strategically using stories created intentionally to elicit behavior change, I believe we can influence better health outcomes and the health care culture change necessary to achieve and sustain zero preventable harm to patients.

Current State

Patient advocates like Helen Haskell, Patty and David Skolnik, and Carole Hemmelgarn have told their stories countless times around the world. Lewis Blackman, the little boy all the little girls wanted to marry; Michael Skolnik, the young nursing student who looked out for everyone, young and old; and the youngest, Alyssa, who was exploring music, the arts, and athletics, still too young to know what she would become—all have given their lives and their stories to improve health care. Their parents, Helen, Patty and David, and Carole, captivate audiences comprised of health care professionals from all walks of the profession, as they put faces to the cost of failure to rescue, to credentialing physicians who should have a license revoked, and to the additional harm and burden of grief that the lack of open, honest communication piles onto patients and families.

Unfortunately, there exists an endless stream of patient stories from which change agents can draw upon to raise awareness around patient harm or a poor patient experience. These stories have helped put the need for change at the forefront of policymakers' agendas. They have increased the awareness of hospital administrators that more needs to be done to prevent harm and invite patients into the conversation about their care.

Since the Institute of Medicine's seminal work published in 1999, "To Err Is Human," called to light the number of patients harmed by medical error, greater attention has been paid to these stories. As a result, a stronger, more intentional call for action has evolved into a demand for health care improvements at a place health care administrators understand: their bottom line. The creation of initiatives such as the

Center for Medicare and Medicaid Service's "Never Events" and the 2008 HCAHPS surveys can at least indirectly be related to the knowledge that flowed from patient stories as this genre was just gaining a stronghold—and is now more frequently shared in the lay press and mainstream media.

On a more local scale, health systems are creating patient and family advisory councils that include a safe place to discuss the internal health system stories that can lead to real and lasting improvements. Programs that reward the reporting of events or the stories of near misses, unsafe conditions, and patient harm—traditionally silenced at the point of error behind the wall of silence—are being elevated so that everyone can learn from these mistakes. Human factors experts have shared that six hundred near-miss events occur before the one that injures a patient, giving health care professionals six hundred chances to learn and revamp operations before anyone is hurt, provided those stories are shared and not buried in fear.

And it is not only patient stories that are being elevated. Providers, often the second victim when harm occurs, are slowly sharing stories that have been buried beneath the same "wall of silence" Rosemary Gibson talks about in her 2003 groundbreaking book by the same name, calling to light the magnitude of patient harm occurring in health care using stories from a cadre of patients and families. Our own title, *Shattering the Wall*, is both a nod and heartfelt gratitude for Rosemary's work in tandem with our "educate the young" premise, and using this same idea of stories as drivers of change, now amplified by the energy and resolve of young health care learners to push them forward. The sharing of health care provider stories, for example, MedStar Health's "Annie's Story," gives caregivers young and old permission to be human, and they serve as reminders that they are not alone. Provider stories also serve as a check and balance of sorts, reminding administrators that there exists equal obligation to create health systems that protect patients and employees alike.

Health care provider safety is a far less talked about concern, as most caregivers are selfless, putting others' well-being before their own. Too many health care professionals know a colleague who chose to end a career, or tragically, their own life, before admitting to struggling in a health care system that fails to prioritize the best interests of those dedicated to caring for others. Many instead are suffering silently at the frontlines, and far too many often choose to harm themselves rather than reach out for help. The Telluride Experience blog also offers a safe space where learners can voice personal experiences they might not otherwise share. Eghosa Isa, TTE Scholar 2016, shares the following:

> I was not going to share this but have been inspired by the courage of others around me ... I also realized that if I want to be part of a movement of change, I cannot demand transparency from a system without being willing to be transparent in my own experiences. This demands a level of vulnerability this is not easy or pretty. However, it becomes completely worth it when finding strength in yourself leads to others finding strength in themselves ...
>
> Graduation day was the best experience of my life. I was finally a doctor! I was going to make a difference. I had such happiness about my life and the difference that I was about to make in the world, and nothing could stop me or my joy. There was nothing that could wipe that smile off of my face and the joy from my heart.
>
> In the first few days of residency, we had a mandatory "emotional harm" meeting. I thought it was nice of them to do—and always a good reminder. It focused on empathy toward the patient, and not losing our empathy when getting in the rhythm of dealing with similar situations and cases over and over again. I loved that they did this. This is something that is so important to remember and necessary to address.
>
> Looking back however, I just wonder, what about my emotional harm? Where are my resources? In this first seven months of my residency experience, two senior attendings committed suicide. I did not know the first, but I certainly knew the second. While there was heartfelt sadness and memorials to honor both, there was

nothing else. No counseling offered to employees, no conversations, no checking in after some days, nothing at all.

Why are these things important? Because over the course of my first year of residency, I have become more depressed than I ever thought possible in my life. Without getting into specifics, the combination of the stress, hours, being made to feel constantly devalued or unimportant, the verbal abuse, the fact that even when I do good it's not good, among many other feelings on an almost daily basis, have caused me to be in a place that I never have been …

So, when these two people directly linked to the place that has been causing my pain decided to take their lives, where was I supposed to turn? Where were my co-residents supposed to turn? Let me offer that I am fine now, but that is no thanks to my system. What about those who don't have the opportunity or courage to turn to where I turned for help?

The human factor's mind-set is so important. The focus on learning instead of punishing is something that should never be overlooked. If I felt more support as such, I know that my experience would have been different. We are all humans and make mistakes. Rarely are errors done with malicious intent. Chances are something in the system failed the employee. I think the mental health of health care workers is vital because I want to know that that person has no other distractions when caring for my loved one.

I am not saying offer full-on counseling sessions for employees because I understand the money factor, but something should be done. This is why I really love the idea that a part of the "Go Team" implements having a conversation with the person who committed the error. A small example, when there are shootings on school campuses, you always hear on the news that counselors will be available to talk with students. Why can't we do the same? I believe that a conversation can change an outcome and save a life on both sides of the health care equation.

Compared to data on the collective mental health of young health care professionals, Isa's personal story conveys the opportunity that exists within the majority of health systems to better reflect on and honor the stories of their own. We can all agree that the delivery of health care can be a stressful profession, but that stress is compounded when providers are left to fend for themselves at times they most need support from their employer and colleagues. This speaks to a larger need for change within health care culture and training in general. Specifically, the unspoken curriculum and the more overt training experience that leads health care professionals to believe they are, or should be, superhuman needs to become more grounded in reality. Health care professionals are not—and never will be—superhuman. The longer this myth persists, the more patients remain at risk as a result.

Lessons Learned

Professional writers and filmmakers know how to design stories so that conflict builds through delicately crafted tense moments and rising stakes that place our beloved hero in jeopardy. We are intentionally drawn into the lives of these characters so that we are invested in their success. Viewers really want Dory to find her family (no spoiler alerts here, just a two-thumbs-up recommendation!). We root for the young wizards that J. K. Rowling gave life to in hopes they will succeed in defeating the evil "He Who Must Not Be Named," and we cry when Dobby dies. We applaud when Atticus Finch gets Tom acquitted, and when the Empire strikes back, we are on board for the ride, showing our allegiance to this mythic coming-of-age story by purchasing millions of dollars' worth of Star Wars figurines. Those miniature Han Solos and Chewbaccas sit poised on our desktop computers, begging our minds to revisit their story world. We pay money to see the story of Romeo and Juliet played out in modern times over and over again because tragic love stories just never get old. Our minds crave the beginning, middle, and end of a good story, and the vicarious roller-coaster ride of emotion that we experience from a safe distance.

Yes, Harper Lee, William Shakespeare, George Lucas, Stephen Spielberg, J. K. Rowling—insert your favorite storyteller here—are all skilled story surgeons. Have you ever wondered how these artists hold us captivated in the worlds they create? Their work inspires storytellers-to-be to create characters, prose, and plays that convince us the seemingly impossible can be achieved and that the status quo can indeed be challenged. The fantasy world they create becomes our reality for the moment. They make us laugh until we cry—or cry until we laugh. Their characters compel us to move, physically and metaphorically, as our hands reflexively shoot up to cover our eyes or to yell, "Look out!" to warn our protagonist that a zombie lurks around the next onscreen corner. We are sent from the theater deep in thought, contemplating how the story just consumed is a reflection of our world—and how that story fits our current worldview. How do these skillful artists increase our heart rates with the precision of an anesthesiologist shooting epinephrine into a vein? How do they force us to revisit how we view the related issues presented as fictional? More importantly for our purposes, how can we learn and apply these skills to shape health care stories that inspire change?

At the very basic level, good health care professionals know that a treatment plan is far better received by a patient if packaged in the right narrative. A friend, colleague, and internal medicine physician, also known as the "Patient Whisperer" in some circles, is known for his ability to spin tales that change the behavior of his patients in ways that often leave colleagues scratching their heads. Somehow, this young physician knew how to frame the medical need to stop smoking after a heart attack into a serene picture of the patient once again, sitting in his fishing boat, smoke- and oxygen-free.

On a larger and more programmatic level, stories are being used as the foundation for culture change in health care organizations in the form of *Good Catch* programs. Good Catch programs reframe near misses, unsafe conditions, and patient harm events in a way that reward health care professionals for sharing stories that often stay hidden by the guilt ridden. Many health care organizations now regularly and publically recognize the frontline and C-suite staff who share a Good Catch story from within the organization.

MedStar Health is an excellent example of a large health system employing an entire Good Catch program, which revolves around a weekly "Monday Good Catch" mailing that highlights the patient safety and high reliability work going on across ten hospitals and more than one hundred outpatient facilities, in a system comprised of thirty thousand associates and six thousand affiliated physicians. Each week, a story that exemplifies a Good Catch from one of these entities is shared throughout the health system.

The program began as a weekly email sent from David Mayer, TTE founder and VP of MedStar Health Quality and Safety, to a leadership committee of approximately thirty people. As the system became accustomed to showcasing their less-than-finest moments, that weekly mailing grew into an internal email campaign that is now mailed to a formal list of about 2,500 associates and an informal list of at least 7,500. The ongoing next step is to get the program officially sanctioned and sent to all thirty thousand employees.

Every month, the quality and safety team at MedStar Health selects a Monthly Good Catch winner, and C-Suite leaders from the local and system-wide teams travel to the winner's location to surprise and celebrate the monthly winner for his or her commitment to patient safety and high-reliability work. All winners and their managers are then celebrated by the executive leadership at an annual luncheon each spring.

MedStar has also partnered with Sorrel King, CEO of the Josie King Foundation, in the past to present the Josie King Hero Award to a MedStar associate who has gone above and beyond to protect and put the patient first during the delivery of care. This recognition program continues to shape the MedStar culture in a positive fashion, and associates are excited to share in story form what they once were afraid to admit outside the confines of the place in which the learning events occurred.

It takes courage to step outside the normal or existing culture of a workplace or classroom—no matter how dysfunctional it may be. Health care has been operating as a somewhat dysfunctional family with a culture that has not empowered caregivers to be open and honest with themselves or patients and families. As a result, stories from either side of the care equation have not been welcomed. The more tenured health care students become, the more they are exposed to a culture that remains in flux when it comes to taking care of those who care for others. By the time they are residents, the existing cultural landscape is in full view, and those who attend TTE understand all too well what needs to be changed and how stories have the power to be of service.

A Telluride alumnus from 2013 offers thanks for using the format of story-based learning, along with a suggestion for expansion of the same type of learning at any health system level:

> I would like to offer my immense appreciation for the *courage* and sharing of stories by families of patients! We are learning from these experiences that in the review of serious safety events, the event analysis is not complete without the patient or family telling of their story as they are the most important part of the health care team …

One former student who is now a professor also knows and utilizes health care narrative in her own work. She provides an excellent closing quote for this chapter:

> Throughout the week we saw, heard, and felt what stories can do for the way we think, act, and make decisions. But telling stories demands trust, and it also demands humiliation because it exposes our natural limits as human beings, which can be incredibly uncomfortable. However, these moments of discomfort are often some of our most powerful learning tools because we open ourselves up temporarily to the possibility of change and transformation, whether we are the storyteller or the listener … Last week was without a doubt the most stimulating and enlightening learning experience I have ever had, so many thanks to all.

CHAPTER 3

Remembering Lewis
Helen Haskell and Gwen Sherwood

Blog Posts Written by Health Care Learners

The Patient in Snapshots
Erin Bredenberg
August 4, 2014

I am very grateful to have had the opportunity to listen to patients and their families tell their stories at TTE. Hearing Helen Haskell talk about the tragic and preventable death of her son, Lewis Blackman, influenced me deeply. Mr. Blackman died because his providers saw him in snapshots and because they saw only what they expected to see. Ms. Haskell perceived him as he was during his fatal hospital stay—in danger and dying—but her concerns were not enough to override the health care team's erroneous conclusions. Through Ms. Haskell's retelling of the story, Lewis is not simply a patient, not simply a statistic or a mistake, but a young man with hopes and aspirations. Ms. Haskell herself is a grieving mother, someone with the courage to overcome her own anger in order to share a deeply painful story in the hopes of preventing others from having to face the same tragic event.

It is humbling to realize how profoundly providers' decisions can influence the lives of patients and their families. Most providers are well-intentioned and wish to help, not harm, the patients they care for.

Yet without recognizing one's own limits, without making an effort to engage patients and their families, without taking the time to listen to patients' concerns, it is easy to make decisions that will actually hurt a patient. As a future physician, I aspire to be someone who listens—truly listens—to my patients, and someone who recognizes the limits of my own abilities, who seeks out help when needed.

A Mother's Guilt
Laura Schapiro
July 30, 2014

I was deeply disturbed, as most people would and should be, by Lewis Blackman's story. The missed alarming symptoms, the lack of respect for the family, the miscommunication, the strained staff came together in a horrific (all-too-common) perfect storm.

What struck me the most, however, was Helen (Lewis's mother) speaking of the guilt she felt. It honestly made me angry to think that after this nightmare she went through, the completely preventable death of her son at the hands of healers, this mother felt responsible for her son's too-soon departure from this world. She thought it was her fault. I find this atrocious that the medical staff left her alone without any apology or closure and left her footing the blame in her own mind. Can you imagine that burden?

Exceptionalism
Daniella Schocken
July 30, 2014

For those who are aware of the risks inherent in our current system of medical practice, there is a dangerous cognitive trap—a trap that catches many good caregivers who otherwise have only the best intentions and the utmost concern for their patients. That trap is the trap of exceptionalism.

I may still be young, but I realized today that I have already made a critical error in judgment.

A tribal tendency persists among many physicians to protect our own from the accusation of wrongdoing or error, but the motivations underlying that tendency arise from multiple considerations. As a doctor, it is easy to think that if one protects his or her peers, those same peers will be more likely to reciprocally protect him or her. I believe the tendency toward protection goes deeper even than that. We defend our colleagues against accusations of wrongdoing or error because we trust them both implicitly and explicitly on the basis of our shared training and experience as physicians.

However, not every patient who suffers an unsupportable decision pays with his or her life—and thank God for that.

Looking in the Mirror
Amishi Bajaj
June 17, 2014

I was fascinated by Helen's words in, "The Faces of Medical Errors ... From Tears to Transparency: The Story of Lewis Blackman," and I jotted down quotes of hers that were eerily similar to those of my mother regarding my grandmother's case. Helen characterized the hospital as a "system that's operating for its own benefit," much as my mother had wondered aloud why everyone seems to be going through the motions of their duties without caring about the outcomes of each completed action.

By far the most angering to me is that last act: that Helen consistently went unheard each time she could see a decline in Lewis's condition. The helplessness on Helen's face and the exasperation in her tone when she stated, "I am the only person who knew everything that had happened to Lewis," made me feel as though as I was looking in the mirror.

Initial Reflections

The profound reflections from the nursing and medical students who watched the Lewis Blackman video provide evidence of the power of story to create awareness, provoke thought, and impact an individual, or group of individuals, in such a way that it can affect change.

Every day, health care professionals identify, manage, and resolve health situations that are sometimes complex, multifaceted, and often unique to the individual patient and family. While this may all be very routine to the providers, to the patient and family, it can be life-changing in ways that busy caregivers may hardly be aware.

As Bredenberg states in "The Patient in Snapshots" reflection at the start of this chapter, health care providers see patients in brief pictures that are only part of the fabric of the patients' lives. This snapshot view may be at the heart of the medical harm epidemic that now places medical error as the third-leading cause of death (Makary & Daniel, 2016).

This reality is also reflected in Bajaj's story "Looking in the Mirror," which comments on the idea put forth in the Lewis Blackman video that health care is a system that operates for itself. Much has been explored about the fragmentation of the health care system and the lack of attention given to the patient experience. Hospitals have been likened to the military or to a prison system where those who enter check their personal identities and belongings at the door and enter a world in which all are, supposedly, treated equally impersonally. It is the way the work gets done. But is it right? Is it effective?

Lewis's story highlights a fault in our current health professions education system that allows inexperienced physicians, many still in the early stages of their education and training, to be primary decision-makers. Too much of the time, they must take action without appropriate backup or support from more experienced or senior staff. To further complicate the situation, these inexperienced providers are too often afraid to speak up or reach out for help when they are unsure of a diagnosis or treatment decision that they must make.

The learners' reflections expose these gaps in our system as contributing factors in Lewis's death. An inexperienced team was in charge and felt intimidated about asking for help when Lewis's condition changed. There appeared to be a lack of focus on Lewis and his changing condition and more of an emphasis on simply completing tasks. This was evident when deteriorating vital signs were simply charted without consideration that this critical information may actually signify a change in the patient's condition. There was no apparent effort to act on the information that was obtained. Failure to act is as much an error as taking the wrong action. Failure to act in Lewis's case also included a failure on the part of the staff to listen to or include the family—or even the patient—in the assessment of his condition or decisions about his care.

We all have to ask ourselves why health care professionals fail to follow evidence-based procedures and put patients at risk. How can caregivers fail to recognize the connection between their actions and their patients' outcomes? If a caregiver does not wash his/her hands, what risk is being passed onward to the patient? If s/he fails to check the potential dangers of a medication, how is s/he placing the patient at risk? In the rush to deliver efficient care, how do we shortchange our goal of effective care? We have to reframe safety from both individual and system perspectives if we are to have transformation.

Change begins with learning the art of listening to the patient and family and to colleagues who may provide valuable perspectives in making decisions. In Lewis's case, hearing multiple perspectives— mother, nurse, resident, pharmacist, physician—could have changed the outcome. All could have shared information as a team. Where was even one assertive voice with the courage to speak up?

Historical Background

Decades of paternalistic health care have conditioned patients and families to have a passive role. Health care information was tightly held by health professionals with selective sharing with the patient and family. The physician held a sacred space and was the undisputed leader of a patient's care. As nurses developed their own professional roles and nursing education was significantly updated to reflect a growing role in the health care team, conflict often arose between nurses and physicians about roles and responsibilities, especially about who made decisions. Nursing education still lagged in teaching interprofessional communication and negotiation as nurses' roles expanded in light of the growing complexity of patient care. Next to families, nurses are the constant in patients' health care experiences. As such, they have critical information to contribute to the team, yet in the past, they often did not have the assertiveness to speak up. Cultivating a culture that redirects authoritarian decision-making to shared decision-making—both between families and professionals and among the professionals on the health care team—helps avoid deference to position and authority.

Traditional methods for handling health care mistakes were based on punitive models in which the person at fault was identified and blamed. Just culture can change that paradigm by investigating the pathway of decisions that led to an adverse event or near miss. This approach can prompt redesigns in the system to prevent the same mistake from happening again. When health care providers focus only on tasks, without seeing the connection to actions before and after their own actions, they put patients at risk. Root-cause analyses help connect actions and consequences and show the connection between actions of the entire health care team; if one individual overlooks something or fails to report a finding, then decisions made later can be affected by omission.

We are all fallible humans regardless of role or skill level. This recognition demands that we recognize the role of systems in improving safety outcomes. Errors that lead to patient harm can happen to anyone, anywhere in the system. Operational processes can be redesigned to minimize risk and prevent mistakes before they happen. In that way, individuals and systems share accountability and commit together to monitor their work to alleviate harm.

Current State

To change the spectrum of error prevention and disclosure requires multiple changes among the system, individual providers, and patients. Advocates for the Planetree (2009) health care philosophy argue that health care could be implemented another way by focusing on patient-centered care. Indeed, the Institute of Medicine identified patient-centered care as one of the six essential domains for measuring quality of health care; it is essential for best care. The IOM went further. If we are to achieve these six aims, we must rethink health professions' education, and they identified the competencies for all health care professionals if we are to improve patient care outcomes and develop a safety culture. And, going further, teamwork and collaboration, other essential competencies, are based on inclusion of the patient and family as active team members. A team implies that all are included in conversations about care in order to have shared decision-making. Collaboration is a way of working together toward a common mission that seeks mutually satisfying resolutions. Only the family has more than a snapshot of the patient; their knowledge and input into care planning decisions is essential if the team is to make informed decisions.

Improving safety is synergistic and involves commitment from all of us. Respecting patients and helping them become engaged, knowledgeable, empowered, and active in their health care management results in better outcomes. Developing mechanisms to improve the flow of communication in a language that is understood and appreciated by all can help patients and their families become full participants in shared decision-making. Active participation as a team member is an extension of effective communication between providers, patients, and families. It means that all parties work to establish mutual goals, monitor performance, share information, and provide feedback. This shared contribution provides more than a snapshot to help place the patient's experience in the fuller context of their life.

The Planetree model demonstrates ways that care can be organized around patient needs and desires to create a personalized, humanized, and demystified health care experience (2009). Daily operations are reframed to focus on the patient and family rather than based only on the system needs. Hospitals implement processes and structures that inspire and enable caregivers to deliver patient- and family-centered care, *and* they help provide the preparation to ensure providers have the expertise, tools, and support to personalize the patient experience.

To alleviate the feelings of powerlessness patients and families often experience during hospitalization, new strategies are being implemented to improve patient-centered care and teamwork. Family-centered rounds take place at the bedside and include the patient and family in reviewing patient progress, setting daily care goals, and clarifying questions. Family meetings with providers involved in care are another mechanism to coordinate care, hear concerns, and improve communication.

One safety valve for patients resulted at least partly from the Lewis Blackman case. After Lewis's death, his family and supporters, including the hospital where he died, worked to pass a law requiring an early patient-activated rapid-response system. The Lewis Blackman Patient Safety Act, which included a requirement for hospitals to have an emergency mechanism that could be triggered by patients, became law in Lewis's home state of South Carolina five years after his death. Patient-activated rapid response gives families access to the same rapid-response teams that can be initiated by providers when they see signs, such as difficulty breathing or a change in heart rate or rhythm, indicating that a patient's condition is deteriorating. Rapid-response teams and other emergency systems like roving nurses who can round proactively on high-risk patients have done much to reduce the number of preventable deaths like Lewis's.

Handing over a patient during transitions in care also presents the risk of missed information, miscommunication, and omissions in care. To alleviate these concerns, handover between shifts may be conducted at the bedside with the patient to increase transparency and communication. Every care interaction should be anchored in a respectful partnership, anticipating and responding to patient and family needs for physical and emotional comfort and for information and education needs. Respecting and responding to each individual is an important part of the patient experience and is a critical component of patient safety.

Creating systems in which safety culture is embedded in mission, operations, and caregiver actions and communications requires a new focus on education and training. The aforementioned competencies developed in 2003 by the IOM and embraced by all health care educators provide a road map toward change and improvement (Cronenwett et al, 2007). The knowledge, skills, and attitudes that make up these competencies are an assurance that together health professionals can recognize the deficiencies in our current systems and commit to recalibrating our daily interactions among fellow caregivers and more particularly with patients so that, in time, we truly can witness transformed systems.

Lessons Learned

There are important lessons in these reflections from nursing and medical students. What we can learn from stories like Lewis's is that patients and families want to be listened to, to be respected, to be cared for gently, to have their caregivers communicate in a clear and accountable manner, and to be given honest and timely information that enables them to make sound decisions (Koloroutis, 2004). As a means of facilitating this, all providers should share the goal of increased self-awareness of the care they are providing and increased situational awareness of what is happening around them. How do we impart the skill to provide this kind of desired care? How do we raise providers' awareness of each patient's uniqueness and put together a holistic view that is more than a series of snapshots?

We asked the question earlier why otherwise competent health providers take shortcuts that jeopardize patient safety. We need to think about how health care professionals can increase awareness and work to alleviate the risks and gaps in health care. We must promote the communication skills that enable individuals—including patients and families—to speak up, address concerns, and hold difficult conversations in a way that gets attention but does not jeopardize the team's effectiveness. We must teach all parties to use critical language to convey the information that all team members, whether patients or providers, need to have in order to function.

Changing attitudes and actions can be daunting. Using stories is one way to surface hidden barriers to change. The personal dimensions of story help reveal multiple viewpoints, challenge assumptions, and guide interpretation. The elements of story demonstrate dynamic real-life applications and provide learners with a vehicle to engage in reflective practice and examine their own attitudes and behaviors (Sherwood & Horton Deutsch, 2015).

Story, as narrative pedagogy, serves as a powerful change agent by applying experiential learning theory to developing a new mind-set for quality and safety. Narrative pedagogy helps learners work within a concrete situation (the story) to apply what they know from previous situations. It also provides a context for applying new knowledge. Narrative medicine is an emerging area of improving health care by sharing meaningful experiences and bringing the patient into the caregiving circle. Some providers ask the patient for a story to include in the patient record in order to expand the perception of the patient beyond the set of signs and symptoms that are the reason they are seeking health care. By seeing a personal dimension of the patient's life, caregivers can see the situation within the context of the patient's life—and provide more holistic care.

In using story and narrative, some questions to ask are:

- What stands out?
- What is the most important action to take? Why? What are the alternatives?
- What assumptions am I making that might influence my response?

Reflective practice is based on inquiry, the art of asking questions to critically analyze experience, make sense of practice, and consider contradictions. To improve safety outcomes, story can help prepare future practitioners with effective team communication. Story is an evidence-based interactive teaching method that integrates experiential learning, reflective practice, and change. By reflecting on stories and cases from practice, caregivers can recognize both gaps and examples of the integration of the six IOM competencies. The story of Lewis Blackman provides an analysis of these competencies to illustrate gaps that led to the cascade of events. Recognition of any of these competencies could have been the trigger to stop the action, have a team huddle, and get everyone, including the family, on the same page so that critical signs and symptoms were seen in their entirety rather than as a snapshot.

How do we help providers achieve this transformative approach? There are many curricular guides, such as TeamSTEPPS, an evidence-based curriculum that helps teach communication skills for all types of providers. These skills are critical for shared information among providers. They are equally useful for patients to help them learn ways to communicate with their caregivers.

Summary

The reflective stories from each of these learners offer a lesson in change. Reflecting on experience helps reframe events and integrate knowledge and experience to help reshape the choices we make in the future. Learning from stories allows us to delve into multiple perspectives to increase both self-awareness and situation monitoring and to consider our own actions within the overall context of what is happening. Health care systems must change; systems rely on individuals to develop the competencies identified by the IOM for safety culture, but systems must also be accountable in supporting their workers and patients in safety behaviors. Health professions' education must be designed to prepare graduates with these competencies so that, in time, we have new workforce approaches. The skills are a starting point to help all communicate effectively. It is up to each of us to embrace this vision of health care; our lives depend on it.

CHAPTER 4

The Broken System
Paul Levy and Farzana Mohamed

Blog Posts Written by Health Care Learners

Bad Person or Bad System?
Aubrey Samost
August 4, 2013

I am a system engineering graduate student, and I firmly believe that the vast majority of bad outcomes in health care is due to good people working in bad systems. However, today when watching the story of Michael Skolnik's death after three years of complications from neurosurgery, I felt like I had just seen one of the rare cases of a truly bad person in the health care system.

For those unfamiliar with the case, here is the two-minute synopsis. A previously healthy twenty-something-year-old male presented to the emergency room after having a syncopal episode. A head CT shows what may be a colloid cyst with no evidence of increased intracranial pressure. An MRI is done and may show the same colloid cyst. Michael and his parents go to see a neurosurgeon who immediately admits him to the neuro ICU. He gets the family to sign a consent form that they barely understand and places a bedside VP shunt to drain the possible excess CSF. Next, the neurosurgeon told the family that Michael needed to have the cyst removed. He said the procedure was small and glossed over any possible complications.

The parents felt they needed more time before signing the consent form for the procedure, especially after feeling like they had been deceived with the last procedure. Later that day, the surgeon returned, and, finding Michael alone, had Michael sign the consent form despite the heavy doses of opioids he was on. The next day, the surgery goes ahead with terrible results. Michael suffered severe brain damage and had nearly every possible complication, none of which the family was prepared for because of the terrible informed consent process. After three years, Michael finally died of these complications from a surgery that it turns out he may never have needed.

As the story unfolded, it felt like the neurosurgeon constantly placed his own needs ahead of those of the patient. I was so angry that he seemed to force the procedure on the family and patient without mentioning any alternatives.

As we discussed this, I realized I wasn't alone in my anger. One of my colleagues pointed out that it practically felt like murder what had happened to poor Michael. As I was listening to these comments and reflecting on my own, my blood pressure slowly lowered. Then, the systems engineer in me started to speak up.

What other factors may have led him to mess up this process so badly? As with any complex system, the answer is multifactorial and more complicated than it initially seems.

- Financial incentives are misaligned: insurance pays you to do a procedure, not to advise the family and patient to not undergo said procedure. The need to get paid could certainly have biased this surgeon into pushing strongly for the procedure.
- Time constraints: The surgeon was most likely in clinic or the OR during normal business hours when Michael's family was visiting. After the surgeon was done in the OR, he could come up to see Michael and get the consent form signed, but that did not align with family visiting hours. Therefore, the system could have acted against him being able to give the family the opportunity

to go through the informed consent process and instead forced him to get Michael to sign it alone.

- Culture: Many of my colleagues remarked after seeing this documentary that they rarely or never saw an informed consent properly done. What the neurosurgeon did here was just another example of normalized deviance. If everyone else in the hospital was signing consent forms this way, is it any surprise that he did?
- Administrative pressures: Perhaps this was the only full-time neurosurgeon in this hospital, and he was under a lot of pressure from administration to not shunt business to their local competitors. This pressure could make him feel unable to turn away a case even if he did not feel totally comfortable doing the operation.

Overall, I have no idea if any of those above system ideas are correct or played any role in this accident. However, by the time I finished thinking through these theories, I felt that we as a class had been missing the most important question. If we want to prevent informed consent failures in the future, we need to ask why he failed to obtain a true informed consent. And when we answer this question, we need to consider the possibility that any neurosurgeon in the same position might have reacted that way because of the pressures the system exerted on him or her. Only then can we change the system to prevent a well-intentioned young surgeon from falling into the same trap and hurting a vulnerable patient and family.

Initial Reflections

This blog post from TTE is characterized by the anger and frustrations that often accompany uncomfortable revelations about one's chosen field, especially for young doctors. After all, they have chosen to devote their lives to alleviating human suffering caused by disease. They studied hard, competed to get into medical school and training programs, and have assiduously followed the rules to get where they are. They are ready to begin a lifetime of service.

And then, almost cruelly, TTE exposes them to uncomfortable aspects of their education to date, their profession, and the work environment in which they will find themselves for many years. Their review of these dysfunctional patterns leads first to dismay and anger.

Samost makes the situation personal with regard to the doctor in the Michael Skolnik story:

> I was so angry that [the neurosurgeon] seemed to force the procedure on the family and patient and never mentioned alternatives. Maybe his motivations were financial, or maybe he just felt he needed the practice with this procedure. Perhaps he wasn't truly evil in intention but just had terrible clinical judgment. Regardless of which of the above might have been the truth—evil intentions or incompetent medicine—my blood pressure was surely elevated by the end of this film because I was so angry.

Another learner offered a deeply felt frustration with the medical education system:

> I am struck by the lack of education in our institutions. Today's lessons were profoundly important and informative, yet our schools do not have room for things in their education such as human factor engineering and negotiations. They don't even make time for true team building!

Such reactions by our TTE learners are a necessary and healthy first step—coming as they do out of a sense of betrayal—but to be productive, anger must be channeled. For her part, Samost offers an analysis of what might have produced the behaviors they witnessed and with initial thoughts about remedies.

Samost concludes: "As with any complex system, the answer is multifactorial and more complicated than it initially seems."

She offers hypotheses of four possible systemic flaws: misaligned financial incentives; time constraints; a culture of improperly executed, informed consents; and administrative pressures.

She modestly notes:

> I haven't got the facts in the case to support these; they
> are really just my own theories and possible explanations,
> but it forced me to think beyond my initial gut reaction
> of blaming the surgeon.
>
> If we want to prevent informed consent failures in
> the future, we need to ask why he failed to obtain a true
> informed consent.

This learner attributes the issues she has discussed to a fundamental lack of knowledge on the part of clinicians, with the simple realization that "imparting knowledge at an earlier stage may prove to be beneficial in the short and long term."

It is not the faculty's expectation that, on the basis of a single class or even several days of exposure to these issues, students will offer fully formed solutions to long-standing systemic problems in the health care world. Our pedagogical approach, rather, is to first build awareness of some of those problems and to train our students to see them when they get back to their schools and clinical settings. Only after learning to see these pervasive issues can one build a coalition for change and engage in corrective action.

A Failure to See

Over the years, we've had the privilege of visiting and advising clinicians and administrators in hospitals throughout the world on the topics of clinical quality and safety improvement, transparency, and leadership. One of the things we've learned is that the institutional organization of health care in a country or state is not determinative of progress in these arenas. Whether a single-payer, nationalized system like the United Kingdom or a multi-payer, mixed private and public system like the United States, or anywhere in between, there are common aspects of broken systems. Likewise, different payment approaches—fee for service, bundled, or capitated—are not correlated with the quality and safety of care delivered.

No, what underlies many of the systemic problems faced throughout the world is the fact that the approach to medical education is to teach doctors that the delivery of care is a series of tasks and that their job is to apply their knowledge, skill, experience, and intuition to the care of each patient. They are also taught that they must not make mistakes. Compounding the problem is that doctors are often not taught the essentials of communication, interpersonal skills, and teamwork. Finally, there is virtually no exposure to cognitive biases, human factors engineering, and the essentials of high reliability organizations.

So, our clinicians enter the highly complex world of clinical care with only partial training. They naturally analyze issues from the limited perspectives they have been taught, and they are oblivious to other perspectives that could result in higher-quality, safer care.

As we visit hospitals around the world, our first step is to ask clinicians to shadow other members of the medical and nonmedical staff—nurses, phlebotomists, transporters, housekeepers, supply distribution folks— and observe them in action. We ask our learners to quietly observe and write down the things they see that impede the staff member from doing his or her job. We ask them to also observe how the staff person works around the impediment.

The experience is revelatory. We often hear comments like, "I don't know how she gets through the day. We make it so difficult for her to get her job done."

Our purpose in this exercise is to train people to see problems in the workplace. We also make the point that every workaround invented by a staff member does not solve the underlying systemic problem and has the potential to introduce waste, inefficiency, and quality and safety problems into the clinical workflow.

The clinicians with whom we deal in the world's hospitals are anxious to jump right in and offer solutions to the problems they have witnessed. That is a natural reaction for highly motivated, task-oriented medical professionals, but it is a misplaced tendency.

After all, our newly sighted observers have seen only a short slice of life of those they have observed. They do not have a sufficient context for what is going wrong, and they do not know enough to intervene without creating unintended consequences.

The exercise, in short, is simply designed to help people learn to see.

Our hope is that they will then start to see things in their own clinical practice areas and will begin to take action.

Coalition for Action

Learning to see is a good first step, but it can only lead to frustration if our clinicians are not also versed in how to design and carry out clinical process improvement. In recent years, instruction in such areas has been made available for practicing doctors, but it remains remarkably absent in undergraduate medical education and graduate training programs. Indeed, that absence is part of what prompted the creation of TTE in the first place.

Nonetheless, the clinical leadership in a number of hospitals around the world has found it possible to change the culture of their broken systems. The approaches used have the following general set of characteristics:

1. The leadership of the hospital is firmly committed to cultural change and is personally and visibly engaged in supporting such change.
2. The organization formally adopts some variant of a "just culture." When errors occur, they are reviewed with an eye toward understanding what systemic problem might have led to the error. The organization is "hard on the problem and soft on the people," recognizing that if one well-intentioned staff member made an error, it is likely that it could occur again. (Note: A just culture does not suggest a free ride for reckless or negligent behavior.)
3. People on the frontline are expected, encouraged, and empowered to call out problems they see in their workplace. When they do, they are treated pleasantly and respectfully by their supervisors. The supervisors, in turn, draw together the resources needed to analyze the underlying cause of the identified problems.
4. A multidisciplinary team designs an experiment to remediate a problem identified by the frontline staff. The experiment and the process around it are analogous to the scientific method.

The current state is noted, a desired future state is envisioned, metrics are determined to evaluate progress toward the desired future state, and a carefully planned and executed improvement plan is tested. The plan is evaluated with reference to the desired outcome. If the redesigned process is successful, the story of the improvement is promulgated to other departments, floors, and units in the hospital. If it is not successful, the team tries again.

Hospitals and other health care settings that employ these general principles of process improvement are truly learning organizations. This is a concept envisioned by Donald Schön in 1973, when he said that a learning organization is one that is "capable of bringing about its own transformation." In other words, rather than relying on external forces like regulatory agencies to improve quality and safety, the organization is engaged in its own self-directed journey toward constant improvement. They understand that there is no virtue in benchmarking themselves to substandard industry norms. Such hospitals are characterized by intellectual modesty, always questioning their own policies, procedures, and approaches. They seek advice and information from others and share what they know.

Finally, such institutions are transparent with themselves and others with regard to their progress and their flaws. They understand that the major societal and strategic imperative of transparency is to provide creative tension within hospitals so that they hold themselves accountable to the standard of care in which they believe. This accountability is what drives doctors, nurses, and administrators to seek constant improvements in the quality and safety of patient care.

Such transparency is also an important part of the medical education process itself, helping the next generation of doctors and nurses internalize the high purpose of clinical process improvement and high-reliability organizations. This result was described well in a personal note to the authors by a medical resident several years after her training was completed. She reflected back on the importance of transparency of the hospital in which she had done her training, one in which clinical outcome metrics (like central-line infection rates) were regularly published by her hospital. Here's what she said:

For me, a trainee at the time, the most important effect was that it underlined a shared sense of mission and purpose around quality improvement. The absence of a sense of purpose of this kind is toxic.

What About Me?

We close this discussion with the most common question received from dedicated young doctors, nurses, and other health care professionals as we travel the world:

> I believe in the framework and the approaches you have discussed with regard to building and sustaining clinical process improvement and high-reliability organizations, and I want to practice in that fashion. But the senior leadership of my hospital (or health system) gives no evidence that they are willing to set that kind of direction. What can I do to make it happen?

It is not easy to lead cultural change from within an organization. The forces of stasis and complacency are powerful and pervasive. Creativity and initiative are often stifled, sometimes assertively and sometimes passively. Indeed, there can be career-limiting risks for young professionals who act outside of the behavioral norms of such organizations.

But the lesson on which we would like to leave this chapter is that it is possible for a few well-intentioned and thoughtful people to achieve local changes in behavior and approach, even in otherwise stifling environments. The key is to think small. Rather than trying to achieve a global change in attitudes in your hospital, pick a small problem to work on. Find a few like-minded people who are interested in forming an informal coalition of the willing. Focus on some clinical or administrative process that has impeded people's ability to do their work well and efficiently. With your colleagues, design a small experiment and test it out. In so doing, do not create large expectations of success; instead, be honest with your small group that this will just be a minor change in workflow. When it succeeds, enjoy the success with your team. And then say, "Is there something else we could work on?"

Proceed slowly and methodically within your floor, your unit, or your area. Let the successes of your experiments create their own momentum and sense of accomplishment and teamwork. If that's all you have accomplished, you will have done some good for your colleagues and likely for patients and families in your area. You may not get recognition or acknowledgement of your efforts by your supervisors or other members of the leadership team, but you will have the satisfaction of having made things better.

It might also come to pass that the good work in your area becomes noticed by those higher up in the organization. If so, they might recognize it and perhaps even learn from your approaches. In such a manner, you may help infuse the process-improvement approach more widely throughout your organization.

CHAPTER 5

A Path without Heart
Tim McDonald

Current State of the Problem

> Before you embark on any path, ask the question: does this path
> have a heart? If the answer is no, you will know it, and then you
> must choose another path. If you feel you should not follow it, you
> must not stay with it under any conditions. A path without a heart is
> never enjoyable; it weakens you and will make you curse your life.
> —Carlos Castaneda, *The Teachings of Don Juan: A Yaqui Way of Knowledge*

In 2002, my anesthesia team had an opportunity to evaluate a patient
on the day prior to an elective surgical procedure. As part of her pre-
anesthesia workup, the complete blood count (CBC) laboratory tests
revealed an extremely low white blood cell count (WBC), albeit not
critically low. The following day, the patient, unaware of the abnormal
tests, arrived early for her procedure. My anesthesia team, along with
the surgical team, failed to note the abnormal test results. The surgery
proceeded on schedule.

During the surgical procedure, the patient lost a significant amount
of blood, but the team did not transfuse her. On the first postoperative
day, the surgical service repeated the CBC. The repeat CBC obtained
on the morning after surgery demonstrated a critically low WBC. This
time, because of the criticality of the result, laboratory personnel called

the unit where the patient was recovering. The lab personnel recorded in the medical record that they had spoken to a nurse named "Mary RN" who was caring for the patient on the unit and relayed the critical test results. The doctor caring for this patient did not act upon these critical test results. There is no evidence the doctor was ever aware of the critically low WBC. The service discharged the patient from the hospital on postoperative day two. They made no plans to follow up on the abnormal test result.

Six weeks after discharge, the patient sought emergency medical services at another hospital not far from ours. The outside hospital treated the patient for sepsis, but they were unsuccessful. Shortly after hospitalization, the patient died as a result of sepsis secondary to a treatable leukemia. From the available data, it seemed this treatable leukemia would have been diagnosed with proper follow-up of the original abnormal WBC at our hospital. News of the patient's unfortunate and preventable death was communicated to members of our hospital leadership almost immediately. Hospital leadership shared those results to the teams responsible for her care shortly thereafter.

A cursory investigation of our care revealed multiple breakdowns in communication that led to her death. The medical team reached out to the leadership of our hospital to pursue the possibility of reaching out to the family of the patient to describe what had happened and express our remorse. We were instructed that contact with the patient's family should not occur. We did not violate those instructions.

After a few months, the plaintiff's attorney for the family filed a lawsuit against our organization and all the members of the medical team who treated the patient. Over the course of the following several years, the defense attorneys denied inappropriate care and defended the organization and the care providers from the allegations of negligence. Our organization paid those attorneys hundreds of thousands of dollars for their defense work. Eventually, years later, the organization settled the case for millions of dollars. No emotional first aid was ever provided to the patient's family or the caregivers involved in this case. Little was learned from the event.

In the words of Carlos Castaneda, we had found ourselves "on a path without heart." Despite public calls for honesty and disclosure and evidence of the value of transparency within other prominent health care organizations (Boothman, Blackwell, Campbell, Commiskey, & Anderson, 2009), the necessary cultural transformation has not yet been realized. Many physicians continue to maintain paternalistic attitudes and behaviors toward the communication of mistakes and other uncomfortable, bad news to patients. A recently published peer-reviewed journal reveals that a significant percentage of physicians admit to their dishonesty with patients and their families (Mello, Boothman, McDonald, Driver, Lembitz, Bouwmeester, Dunlap, & Gallagher, 2014).

The perceived barriers to transparency in health care are many in number (Mello et al., 2014). Researchers have identified many of these barriers and some of the solutions. Public policy and societal approaches, including those of the legal community, continue to pose barriers to transparency. The National Practitioner Data Bank and most state licensing boards continue to require medical malpractice insurance carriers to report their insured whenever a payment is made on their behalf. Of note, payment of money, independent of any behavioral or "just culture" assessment, is the only trigger for reporting. Perceived fears of the outcome of the Data Bank and Licensing Board reporting play a significant role in limiting the physician adoption of communication and resolution or full-disclosure processes following patient harm.

In addition to the regulatory barriers, other impediments to open and honest communication include a persistent and ongoing list:

- culture of shame and blame and opaqueness
- fear of litigation in general
- lack of standardized process
- absence of institutional support
- financial considerations
- failure to embed the necessary behaviors in young learners

The "wall of silence" continues to wreak havoc and impede progress and improvement within the health care system. The failure to provide immediate, open, and honest communication and emotional first aid to patient, families, and care providers following harm or inappropriate outcomes exacts a significant toll on patients and care professionals.

The guilt, lack of learning, and the "hidden curriculum" that our learners experience in clinical environments limits the learners' growth, maturation, and learning associated without the necessary conditions predicated for achieving high reliability. It also inhibits their ability to experience joy and meaning in their work.

Learner Reflections on the Lack of Transparency

At TTE, we asked the learners to post their reflections on the topic of transparency—or lack thereof. Some of the more poignant reflections are provided here, with comments. It is evident from these reflections that many of the learners find themselves on paths without as much heart as they desire in their early careers in health care.

Christian Gausvik (2014) authored the blog post "Transparency Might Kill the Deal, but Not the Patient." He recounted the "lack of transparency" experience at an assisted-living facility where he worked during his undergraduate years:

> Just having completed my first year of medical school, my clinical experience is limited, but my exposure to the world of health care is not. Having spent the four years of my undergraduate career working with the geriatric population, I had great exposure.

The learner went on to explain that the sales team at the facility where he worked often spouted words of assurance to families about the excellence of the services their family member would receive. To his dismay, he learned otherwise. "Over time, I came to know that regardless of their legality, the promises the salesmen/women recited held little truth. The deals were not transparent, and at a time when families desperately needed honesty."

The learner goes on to describe:

> Their false sales pitches flashed through my mind the day Mrs. A was wheeled out on a stretcher with a significant cut to the right side of her face. I knew someone in my

department had recently paged a nurse aide to walk her to her room after dinner—something that had been promised the family in one of those sales tours. No one ever came, and so Mrs. A walked herself unsteadily until she tripped. In reality, the facility was never staffed to be able to walk every resident back to his or her room. Despite the frequent promise, it was rarely done ... I started wondering what the family would be told. Would they get the truth, or as is often the case with geriatrics, would the system's shortfall be blamed on the patient's age and mental deficiencies?

Furthermore, the learner stated:

This patient did not belong in the facility I worked in, and this would have been avoided had the family become full disclosure and this patient been placed in full nursing care. Financial pressures to sign leases pushed the sales department to say whatever they had to in order to close the deal, and families were comforted by the show they put on. They were anything but transparent, knowing full well the two staff nurse aides on each shift could never give the level of care being promised to ninety-five residents. Mrs. A spent two days in the hospital because of this lack of full honesty and disclosure—a stay that could easily have led to any of a host of hospital-acquired diseases, such as delirium, functional decline, or a fall—all common in the elderly. It's not hospital transparency exactly, but it's a story that sticks with me in my medical training and reminds me just how important the things we say are.

Through TTE, the learners engage in reflective learning exercises throughout the several days of immersion. Following the reflections, the learners propose and implement projects designed to address and provide solutions to the problems they have identified and analyzed during the immersion. In this case, the learner identified the opportunities around holding assisted-living facility administrators accountable for the promises and assurances they make to prospective patients and families.

Erin Bredenberg (2014), TTE learner, blogged about the "power of knowing what happened … [or not"]. To describe the lasting impact of a patient harm event after which no caregiver support was provided, she begins by saying:

> The past few days [at TTE], a ghost from the not-too-distant past has been visiting. Someone I didn't know was still with me. Someone whose impact I didn't fully appreciate at the time.
>
> The patient was a man who seemed much younger than his thirty-odd years, in part due to some mild cognitive impairment. He had a childlike vulnerability, a fragility that was accentuated by the fact that he was very ill, and nobody could quite figure out why. I was the subintern on the medical service, and he was not my patient, but I met him on rounds, heard the other team members talk in a worried manner about his hospital course, and observed that his frightened and fiercely protective parents never left his bedside.

She goes on to describe subsequent events:

> And then one morning, I arrived at the hospital to find the third-year medical student ashen-faced and shaken and the rest of the team somber. The young man had suddenly died the previous evening during a procedure under anesthesia, and nobody knew why. The student had watched the whole thing; the intern in charge of the man's care had been called to the hospital immediately. Afterward, the two of them together went to tell the parents, who were eagerly awaiting their son's return to the medical floor, that he had not survived.

Poignantly, the leaner provides some insight into the impact of this unfortunate outcome on the care team. With a certain clarity known only by those in the midst of the care processes, the subintern shares the important unanswered questions and concerns of the team.

> For the rest of that month, the medical team talked about the young man constantly. What exactly had gone wrong during the procedure? Should the procedure never have been ordered in the first place? What, if anything, could they have done differently? The attending suspected perhaps something had gone wrong with the anesthesia, though the anesthesiology team insisted that nothing was wrong on their end. Everyone felt guilty. Everyone felt terrible for the family.

There seems to be no indication that the organization or the clinical service had a process in place to support the clinical teams when these types of sad events take place. "Everyone felt terrible," they explained, but no one seemed to have embraced the caregivers in any sort of healing or learning from the events.

> Eventually, an autopsy was done showing that the young man had metastatic cancer and was undoubtedly far more fragile than even the team had suspected. As far as I know, that was the conclusion of the story: the patient was sicker than anyone knew, and he died. The month ended, I moved onto my next rotation, and the residents and other student did the same. I never saw the family again. In fact, I mostly forgot about them.

As often happens to the "second victim," the caregivers move on from the harm event and on to the next patient or the next clinical rotation. The effects of the event on the psyche of the clinician seem to dissipate, but the dissipation is only temporary. At unpredictable times thereafter, much to the detriment of the clinician, the feelings and emotions surrounding the harm event will resurface.

The TTE learner describes the resurfacing of the event:

> The past few days at TTE, I and the other students
> have heard powerful stories from patients about their
> experiences with medical errors. We've heard them talk
> about the shock and grief of dealing with a loved one's
> unexpected injury or death and about the pain of not
> having answers to the question, "What happened?"
>
> I don't know if the patient I met on my subinternship
> died because of a medical error or if his death was truly
> unavoidable. In fact, I'm not sure if anyone knows. I
> also don't know if the family was aware of how many
> unanswered questions the medical team had about the
> conditions surrounding the young man's death. I suspect
> nobody ever told them. In retrospect, I think that it would
> have been therapeutic for both the team and the family
> to open the doors of communication and that it would
> have helped the family immensely to know of the team's
> doubts, fears, concerns, and guilt.

As with all of the learners, this TTE participant pondered solutions to the
problems and opportunities she had identified during her reflections. A
solution the learner identified, one echoed throughout the many years
of the experience, involves the recognition of the importance of creating
processes to "openly, honestly, and thoroughly answering the questions
'What happened?' and 'Why did this happen?' when such an event occurs."

Another important insight obtained during this learner's reflection
included this observation: "Not only can the answers to these questions
help prevent future deaths. They can also be healing for patients,
families, and providers alike." The articulation of this added benefit of
open and honest communication drives home the need for the broad
implementation of programs that provide immediate emotional first aid
to all those impacted by patient harm events—a stated goal of TTE.

Indeed, recent literature suggests the health care industry has much
to learn from aviation in this regard. The difference between the aviation
approaches to supporting those involved in mishaps versus health care

is startling. Following the famous Miracle on the Hudson landing of US Airways Flight #1549, teams of trained specialists reached out to assess and treat all those involved in that flight. Of note, the air traffic controller was given thirty days paid leave because of the emotional impact the event had on him (Stiegler, 2015).

Contrast that response to the standard response in medicine when the surgeons, anesthesiologists, and nurses are expected to proceed uninterrupted even when they are involved in an unexpected death in the operating room. Someday, to do otherwise will be viewed as a "sign of weakness"—another topic discussed during TTE. In addition to the provision of immediate emotional first aid, others have studied ways to help the clinician gain wisdom following unexpected harm. Just a few of those processes that help healing and gaining wisdom include (Plews-Ogan, May, Owens, Ardelt, Shapiro, & Bell, 2016):

- talking about the event
- learning through the event-analysis process
- engaging in process redesign—focused on future prevention

Why We Must Change

In health care, our patients and providers continue to experience too many unnecessary harm events. We continue to fail in the moral imperative to effectively engage and be honest with each other, our patients, and their families when harm does occur. Many believe these two sets of facts— the lack of transparency and the persistently high rate of unnecessary harm events—are causal. Hiding medical mistakes from each other and our patients by engaging in the traditional "deny-and-defend" tactics of the defense bar is anathema to the kind of "learning environment" we need to become a highly reliable industry. The purpose of programs that encourage, support, and provide a rapid response to patient harm, which include open and honest communication and emotional first aid to all involved, is to primarily cultivate a culture of learning and a culture of patient safety.

Another reason we must change and shatter the wall of silence rests with the ongoing struggles to deal with burnout in the health care professions (McDonald, Helmchen, Smith, Centomani, Gunderson, Mayer, & Chamberlin, 2010).

As our TTE learner posted:

> I think that it would have been therapeutic for both the team and the family to open the doors of communication, and that it would have helped the family immensely to know of the team's doubts, fears, concerns, and guilt.

The approach this learner posted is precisely what is needed in health care if we are to restore joy and meaning to the workplace.

These observations further solidify the power and importance of educating the young as they prepare to become the next generation of health care professionals and giving them the tools to engage in transparent conversations—open and honest communication whenever anything unexpected happens during health care delivery processes. If the processes described above were embedded within my organization in the 2002 case described at the beginning of this chapter, we would have responded immediately upon notice of the patient's death. That response would have included reaching out to the patient's family and engaging in open and honest communication with them. This communication would have included an optimal resolution of the event, including appropriate compensation without the need for them to file a lawsuit. This response would have prevented the years of litigation and the unnecessary, painful, and costly depositions that took place. We would have learned from the event. We also would have been able to provide emotional first aid to the clinicians involved in the patient's care.

We would have been on a path with heart.

CHAPTER 6

The Missing Piece
Anne Gunderson and Jessica Gunderson

Blog Posts Written by Health Care Learners

Be Hard on the Problem and Soft on the Person
Elissa Falconer
June 17, 2014

Be hard on the problem and soft on the person.

These words stuck with me after our breakout discussions early this morning. In acknowledging a systemic error, we acknowledge that the situation could have happened to anyone— they just happened to be in the wrong place at the wrong time. In accepting this culture where this would be the norm, we have to break away from our traditional "deny-and-defend" culture.

John Nance mentioned today that our greatest threat is the failure to infuse urgency into the frontline; we must change the way we teach, promoting teamwork. Even in my first year of medical school, we had attempts at teamwork and small groups to promote a team environment. But what was my overall takeaway of what my first year was full of? Studying, studying, studying, and then taking an exam. What do exams harbor? Competition amongst ourselves; we're graded on how few mistakes we make. It breeds a culture of perfection. Sure, we offer exam reviews that are, of course, not mandatory. What if we were required to work in groups through exam reviews—and our professors were present to listen to the questions we had? What message would we change? Would we feel so isolated and feel like perfection was required?

Stuck in the Middle
Christine Thomas
June 16, 2014

> Be the change you wish to see in the world.
> —Gandhi

I would 100 percent complain about the state of patient safety in American health care. I would also 100 percent want to help change this culture, yet I feel stuck in the middle. Stuck in the middle between knowledge and action. It's like I've been given this great insight and passion without an outlet. As a student, I feel very privileged to have access to this information and strategies for how to reduce medical error. I know many of my colleagues do not. However, as we discuss the overhaul that needs to happen within the American health care system, I realize that, as a medical student, the waves I can make more closely resemble the ripples of a small stone thrown into the ocean on an already windy day.

Staying Grounded at Ten Thousand Feet
Caitlin Farrell
June 19, 2014

During our third day in TTE, many of the students and faculty took a traditional six-and-a-half-mile hike up the mountain to a waterfall. This was a completely amazing experience. Standing on top of the rocks, overlooking miles of pine trees, you feel like you are on top of the world. No one else can touch you up there. You can reach up and practically touch the heavens.

Settling back into our studies on patient safety, I realize that the feelings I had while standing ten thousand feet above sea level are not unlike many of those in health care. It is easy to feel as though you are standing at the top of a Colorado mountain, day in and day out. Nothing else, and no one else, is relevant—or can penetrate your thoughts. You, and you alone, can see what is below you. Only you can understand. You look down on the patient and can see things clearly because you are the only one standing from your perspective.

It is very easy to feel this way. I have seen many health care professionals fall into this trap. Patients are secondary to our own thoughts and desires. We are not a small appearance in their lives, but they are visitors in our hospital. They are a tool that allows us to demonstrate our own skills and expertise. They do not run the health care system; they are only subject to it.

How very wrong we all are. It is only when you stand at the top of a Colorado mountain and take a moment to look down, look above, and look beside you that you can clearly see reality.

Patients are not merely objects that enter the health care arena in order to please us. Patients are the heart and soul of health care. They are the reason health care exists. It is this fact that must ground us in reality. We must remember why we all hiked up the mountain. It is not only for ourselves, but for all of those around us, so that we can gain greater perspective together.

Initial Reflections

Falconer's memory of their first year of medical school is an accurate portrayal of the current state of medical education. Students are tested based on the degree of their studying through examinations. Learners are expected to transition from medical school to a hospital environment with this textbook knowledge and perform to perfection by patients and their families who place their trust in these caregivers. According to Derek Blok (2014) in his blog post, "How Much is a Ten-Dollar Bill Worth?":

> The constant pressure to see more patients, do more procedures, and generally be more "efficient" with our time often causes us to jump too quickly to a diagnosis (premature closure). Once we've jumped to a diagnosis, and confidently shared it with others, we are too prone

to cling to it beyond rationality. Our ego gets in the way
of us reevaluating, asking what we may have missed, and
being open to different opinions. All of these things are
detrimental to the health of our patients.

As a result, more errors are caused by the blindness of the professional's
ego, impulsivity, and irrationality.

Therein lies the problem with our medical education today. We teach
students that they need to be right, but we do not teach them *how* to be
right. As Falconer states, we expect students to study hard, pass their
examinations, and be at the top of their classes. This type of learning
does not foster value for patient-centered caregiving or ways in which we
can identify and reduce medical error. We need to teach students how to
prevent errors in clinical practice by focusing on improving the attitudes
and skills of medical students through an experiential curriculum that
goes beyond developing a textbook knowledge base (Halbach, 2005).

TTE is a model of this style of teaching. Medical students and
residents are given the opportunity to communicate through roundtable
discussions, hearing "firsthand accounts of successful implementation of
practices and/or programs to improve patient safety" (Mayer, McDonald,
Doyle & Granzyk, 2011). They are able to listen to cases of medical error
from patient advocates, collaborate in team-building activities, and
create a community of future leaders and champions in patient safety.
This interactive, communication-based curriculum creates awareness,
mindfulness, and compassion for the most important person in any
hospital setting: the patient.

Students who attend TTE are aware of the presence of medical error
in our hospitals; however, not all medical students in the United States
have access to the knowledge about ways to reduce errors. Even if they
have knowledge, their actions to reduce error are lost in the large sea of
health care professionals who were not taught to value the patient first,
recognize the possibility of an error, and prevent an error from occurring.

As Farrell mentions in her blog post, doctors fall into the "trap" that
health care professionals are "secondary only to [their] own thoughts
and desires." In reality, patients should be recognized as "the heart and
soul of health care" and the "reason that health care exists." This is a

fact that *every* health care professional needs to understand, and we can teach it by completely revitalizing our medical education curricula. We need to update our medical curricula in every institution to encompass the lessons and educational structure that TTE exemplifies. By teaching our students a "patient-first" mentality and giving students an outlet to discuss medical error cases, we can increase their comfort and confidence with speaking openly and honestly with fellow health care providers and patients. If we can successfully integrate patient-safety education into our medical curricula, we can make it the standard for any medical or residency training program.

Historical Background

The following passage describes a case study at New York Medical College (Halbach & Sullivan, 2005) where questionnaires were distributed to medical students asking them to evaluate a new patient-safety-oriented curriculum. This study also notes the overwhelmingly positive response of the students to this curriculum change:

> In the years since the release of the Institute of Medicine (IOM) report, "To Err is Human: Building a Safer Health System," in November 1999, there has been a dramatic increase in the number of articles in the medical literature on patient safety and medical error. There has not been the same explosion of publications on educating medical students about these issues although the education of physicians beginning in medical school has been increasingly advocated. From the early 1980s, the literature has included observations of trainees in relation to medical errors, anecdotes involving medical students and residents, and a few interviews and surveys. However, this literature rarely offers specifics on how this teaching should be done. In fact, commenting on progress made and improvements still needed in the years since the IOM report, Timothy Flaherty, MD, chairman of the Board of the National Patient Safety Foundation, noted

in an interview that medical education is an area where patient safety has not made any dramatic improvements.

From 2000 to 2003, third-year medical students at New York Medical College, Valhalla, New York, were required to participate in a new curriculum on patient safety and medical errors during their family medicine clerkships. Five hundred seventy-two students participated in a four-hour curriculum that included interactive discussion, readings, a videotape session with a standardized patient, and a small-group debriefing facilitated by a family physician. Before and after participating in the curriculum, students were asked to complete questionnaires on self-awareness about patient communication and safety. Curriculum evaluations and follow-up surveys were also distributed. Responses to each statement on the before and after questionnaires were compared using the Wilcoxon signed-rank test for matched data.

511 (89 percent) students reported that the opportunity to present an error to a patient increased their confidence about discussing this issue with patients, and 537 (94 percent) students reported that they strongly agreed or agreed that the standardized patient and feedback exercise was a useful learning experience. A total of 535 before and after questionnaires were used in the analysis. A comparison of before and after questionnaire data revealed statistically significant increases in the self-reported awareness of students' strengths and weaknesses in communicating medical errors to patients.

These findings suggest that awareness about patient safety and medical error can be increased and sustained through the use of an experiential curriculum, and the students rated this as a valuable experience.

This study was conducted more than ten years ago, emphasizing the learners' need to incorporate a patient-focused view into their studies. How has medical curricula evolved on a national scale since then?

Current Sate

According to Johns Hopkins University researchers, medical error is now the third-leading cause of death in the United States with more than 250,000 estimated deaths per year (Cha, 2016). This cause of death surpasses respiratory disease, accidents, stroke, Alzheimer's disease, and diabetes. This statistic alone validates the reason why medical curricula needs to continue to evolve and include patient safety education programs. Although patient safety has been increasingly recognized as a key dimension for quality care, few medical schools have modified their undergraduate level curricula to address the necessary educational foundation regarding patient safety competencies. While the lack of these programs is apparent in our education system, there is also an apparent need for them. Improved health care practices and safety begin with improved health professional education.

We need to focus on a shift in culture within our medical schools and our hospital settings by opting for full disclosure. Full disclosure has been shown to benefit patients, health care providers, and the system in which health care is provided. Reports in the literature and actuarial data suggest that aggressive full-disclosure programs can produce greater patient trust and satisfaction and result in fewer malpractice lawsuits. Open disclosure helps health care providers and organizations learn about the problems associated with their processes and systems of health care delivery and prompt improvement in those practices, thereby reducing the potential for harm to subsequent patients. Full disclosure of medical errors related to care is also recommended as a means to achieve safer care for the patient.

The first step in a culture of error disclosure is to educate physicians and other providers. Despite the acknowledgement that full disclosure is essential to professional behavior and the facilitation of patient safety, there are very few opportunities in medical education for learning regarding this critical issue. There is also a lack of information in the literature regarding the constructs required for successful adoption of full-disclosure principles. It is difficult to create appropriate undergraduate clinical learning opportunities when attending physicians, who serve as preceptors, have little or no experience with the disclosure of medical errors, believe they provide optimal quality patient care, and believe they do not make mistakes. Few medical schools have responded to the calls for quality and safety education through introduction of appropriate curricula to prepare students for practice in a transparent health care environment. This critical lack of patient safety education in undergraduate medical training needs to be addressed.

To ensure patient safety, the next generation of physicians must be prepared to recognize potential sources of error in medical practice, to acknowledge their own vulnerability to error, and to engage fully in the process of continuous quality improvement.

Lessons Learned

Here are the reasons why we need to encourage the teaching of patient safety and for medical schools and residency programs to integrate education about patient safety and medical error into their training:

1. The health effect of medical errors on society is huge and merits dedicated time in the curriculum.
2. Academic medicine lags far behind other health care and regulatory bodies such as the JCAHO, the National Quality Forum, and many state governments, and it should be leading efforts to address patient safety problems.
3. Medical schools should address the concerns of patients and the public, many of whom want physicians to handle errors and disclosure differently.
4. Physicians report that more training in how to handle errors is necessary, including ways to constructively heal themselves and colleagues after making an error.
5. There is a need to decrease the emotional and cultural barriers in medicine, to address the "hidden curriculum" in medicine, and to facilitate a change in culture.

Here are lessons and takeaway points for educators to integrate patient safety teaching with the medical curriculum at their institution:

1. Use TTE and medical education models as examples for incorporating multiple methods of patient safety learning. Include opportunities for discussion, guest speakers, training DVDs, simulations, case-based conferences, and team training exercises in addition to textbook examinations.

2. Instill confidence in our nursing and medicine students to be honest and transparent in their professional-to-professional communication and professional-to-patient communication.

3. Focus on establishing competencies in the Four R's of Apology: *recognition, responsibility, regret,* and *remedy.*

4. The next generation of physicians must be prepared to recognize potential sources of error in medical practice, acknowledge their own vulnerability to error, and engage fully in the process of continuous quality improvement

CHAPTER 7

Confessions
Kim Oates

Blog Posts Written by Health Care Learners

Oasis
Evan Skinner
July 30, 2014

The oasis is often recognized as a critical milestone and safe zone along an austere and barren route. To travel deserts without the incidence of a viable oasis was essentially certain death. Sometimes in health care, the landscape and trajectory of our professional careers begins to resemble a sojourn through difficult terrain, hopefully punctuated with spots of salvation and rest.

In our reflection session today, I stated that I desire to renew hope and trust. I stated so as I have begun to become disillusioned with American health care and its potential future state. I am less hopeful at times because I perceive it as a decaying institution that is terribly expensive, inefficient, and dangerous in application. Most frighteningly, as a complicit associate of American health care, I am compelled to look inward and account for my own misguidance in action and the remedial opportunities I never sought.

As a nurse, have I ever participated in an action that ultimately harmed a patient? Yes. As a nurse manager, was it easier to execute disciplinary action at times than to construct a performance-improvement plan? Yes. Have I displayed a hardened and nasty general comportment in my conduct toward others in the hospital from time to time? Yes. Therefore, I have played a role in facilitating the deterioration of a profession I once held dear. As such, I have arrived at event horizon for change. I indicated a need for trust and hope to restore a failing confidence in a system that is so critically necessary to the sustainment of public health and our future as citizens.

I am hopeful tonight that my experiences here at TTE will serve as an oasis for me to renew the foundations of my own practice and to be the change I desire to see in the world, however small that scope may be.

Realizations
Mona Beier
August 3, 2014

I have to say these past few days have been eye-opening, and dare I say, life changing. These talks, stories, and reflections have all made me take a step back and realize what it is all about: our patients. It makes me really sad that I have completely lost sight of that in the very little time I have been in training. There is really no excuse. I could blame exhaustion, long hours, too many patients in too little time, but at the end of the day, there is no excuse for not putting our patients as our number one priority.

Fear, Forgiveness, and Father's Day
Caitlin Farrell
June 16, 2014

Yesterday was Father's Day, 2014. I woke up before everyone else in my room. Rolling out of bed, I padded down the stairs and brewed a cup of much-needed coffee. After pouring a steaming cup, I looked out my window to the inspiring landscape of endless white-capped mountains. This year marks the ninth Father's Day that I have spent without my dad, but the mountains and my purpose this week made me feel as though he were standing there with me, sharing our cup of morning coffee, just as we used to.

I know that despite the growing number of "apology laws" that protect and even mandate physicians to apologize to families after catastrophic events, few physicians actually do apologize. This results in families feeling like the events were their fault. I can say from experience that this is a burden that you can carry with you for years to come.

As I got back to my room and put down my books, I mulled over this conversation in my mind. The death of my father has given me the fuel to pursue medicine and patient safety as my career. It has instilled in me passion, energy, and determination. Yet the one thing that I have not found in the nine years since my father's death is forgiveness. Although I do not hold any one doctor or nurse responsible for the detrimental outcome in my father's care, I have not been able to forgive the team for what happened. I have not been able to go back to that hospital. And, as I sat on my beautiful bed in the mountains, I realized that I also harbored another feeling: fear. Fear of becoming a physician who does not practice mindfulness, who does not partner with my patients, who does not apologize for my mistakes. I am afraid that, despite my best intentions, I will only continue the vicious cycle. A fear that I will allow my patients to feel as though they are "on an island."

I put away my computer, got into bed, and took in the gravity of the day. I am so grateful to be here at TTE among students and faculty who share my passion for patient safety. I could not imagine a more perfect way to spend Father's Day.

Initial Reflections

Oasis

This blog post is a lovely analogy of trudging through the desert, thirsty with the occasional spot of rest.

As well as causing thirst, trudging through a desert can be lonely. We may be trudging through a health care desert, a desert where the technical care may be very good but where the culture is not good.

It can be lonely being the person out of step in a culture that does not always put the patient at the center of every decision. Lonely in a culture that does not involve the patient in discussions and decisions as true partners. Lonely in a culture that perhaps focuses on blame and process rather than on repairing problems and attitudes in the system that may be causing harm to patients. It's lonely, thirsty work. That's why the oasis is so important. It's a place where you can be refreshed, where you can meet other lonely travelers who are facing the same obstacles you face, discuss problems, and share solutions.

We can't help being influenced by the attitudes of the people we work with, particularly our seniors. It's easy to focus on the process rather than the underlying problem and not respect colleagues and patients and denigrate them because that's what others do. That is why it is so important to seek out good role models, role models who can be a beacon for us, an oasis. And that's why it's important for people who have a passion for patient safety to build networks, to talk with each other, to share frustrations as well as solutions, and to learn from each other.

When I think of good attitudes and bad attitudes, I often think of the story of the wise man who told his grandchild that everyone has two animals inside them. One animal is angry, aggressive, resentful, and likely to attack at any moment. The other is friendly, kind, generous in nature, and able to learn from experience and improve. These two animals are having a battle to see who can take control of the person they inhabit.

The anxious grandchild asked, "But, Grandad, which one will win?"

The grandfather gently replied, "The one you feed."

We need the oasis, good role models, and a network of like-minded colleagues and mentors so that we feed the right animal.

Realizations

After listing the reasons why we may not apologize, this blog post hits home: "There is no excuse for not putting our patients as our number one priority." How true—but not always easy.

We find it difficult to have patients question us and challenge us. That just might be because our number one concern is ourselves: our reputations, our expensive and lengthy educations, and our status as doctors. Once the patient is put first and seen as a real person with real needs, we see it differently. We listen to them, stop labeling them, and thank them for their input. Labeling a patient can be dangerous. It may stop us from recognizing serious problems. When I hear of patients being given labels, I think of Carole Hemmelgarn's moving story of a health care tragedy. She said, "Labels belong on jars—not on people."

Fear, Forgiveness, and Father's Day

When health professionals don't acknowledge their errors and don't tell patients the truth, it makes it difficult to forgive them. It makes people lose trust in a system that should be there to help them. It can cause ongoing resentment. You can only forgive when someone admits they made a mistake and acknowledges that their actions may have led to patient harm. Otherwise, forgiveness is difficult.

Forgiveness does not excuse the error. Humans will always make errors. Forgiveness is not incompatible with justice. When we know the person or system contributing to the error expresses remorse, forgiveness becomes possible. And, of course, when we are open about our mistakes, there is a much higher chance of learning from our errors so that processes and systems can be changed to make care safer for others.

Final Reflection

One blogger was asked during a team-building session whether physicians were afraid to make mistakes for fear of litigation and whether it may have hampered open and honest communication.

There was a paradox between entering the profession to heal and not to harm. We recognize that we aren't infallible—and that harm does occur. When this happens, we can be honest and admit our error—or we can hide it. Do the rules change as we grow up? As children, we are taught to own up, take responsibility, and always apologize. It should be the same when we are physicians. A medical practice should be built on honesty, integrity, accountability, and respect and not on fear of being sued.

A strong medical practice should be one where patients can openly say, "Yes, my doctor made a mistake, but he/she took full responsibility and apologized for what happened."

I like the way this blogger pointed out that the rules we learned as children, about apologizing and taking responsibility, should be no different when we grow up. Perhaps that's because I'm a pediatrician and am aware of how important early childhood influences are in shaping behavior later in life. And, of course, the rules don't change. What does change is the way we get better at working around the rules instead of admitting mistakes and apologizing. We get better at making excuses, blaming others (even the patient in some cases), obfuscating, and taking refuge in medical jargon that the patient can't understand.

It's hard to apologize and admit error. It's particularly hard when we put ourselves at the center instead of putting the patient there. It becomes so obvious once the patient is at the center. Litigation is less likely when there is honesty and transparency.

It may take a generation to have a health system where honesty, openness, transparency, and patient-centered care are the accepted norm. You are that generation.

Historical Background

Open disclosure is a relatively new concept in health. In the 1980s, it was recognized that litigation arising from adverse events was rising. In response, in 1987, the Veteran's Hospital in Lexington, Kentucky, embarked on what was then a very radical plan: "To maintain a humanistic, care-giving attitude with those who had been harmed, rather than responding in a defensive and adversarial manner."

Those of us with an interest in patient safety may see this as a perfectly reasonable, obvious plan, but this was almost thirty years ago. It was a concept at odds with the traditional medical view, including medical teaching. That view was to "admit nothing and put nothing in writing."

Looking at this plan through the lens of time helps us realize what a radical, bold concept this was. There must have been opposition, but, to its great credit, the hospital's risk-management committee decided that the hospital had an obligation to reveal all the details of its investigations to patients and family members where care had been compromised by errors or negligence. This included events where the patient or family may not have realized that a mishap had occurred (Woods, 2007).

Sometimes bold initiatives falter and fizz out. This one didn't. And there were clear benefits for the VA hospital as well as for patients. Within ten years, this new practice of apologizing for errors and complications had led to a drop in claims and court cases (Kraman & Hamm, 1999).

Other health systems in the United States followed, with the Joint Commission on the Accreditation of Healthcare Organizations (2001) and the American Hospital Association (2004) adopting a policy for the disclosure of adverse events. The National Health Service in the United Kingdom adopted a policy to implement a "duty of candor" in 2003. Canada and Australia adopted a set of disclosure guidelines in 2008. Open disclosure is now part of health policy across the United States, Canada, United Kingdom, Australia, and New Zealand.

The United States has gone further than other nation by the Joint Commission on Accreditation of Healthcare Organizations (JCAHO) linking compliance with accepted open disclosure standards to hospital accreditation. Compliance with standards is publically available, and funding is linked to compliance with a series of safe practice guidelines. This combination of financial incentives and consumer access to health care provider data has the potential to improve the effectiveness of open-disclosure programs. Some states have adopted apology laws that protect specific information conveyed in open disclosure.

There are various terms for open disclosure and various definitions depending on the jurisdiction or country. Legislation is not uniform across and within nations; most jurisdictions currently protect an admission of regret rather than an admission of liability. What they have in common is that they all emphasize communication with patients and their families after an adverse incident—and the relevant communication must be open and honest.

The main elements of open disclosure in the majority of jurisdictions and nations that have embraced it include:

- prompt provision of care to treat the problem caused by the adverse event and protect the patient from further harm
- acknowledge that an adverse event has occurred
- inform the patient and family as soon as possible

- express regret for what has occurred, including an apology if appropriate
- provide information to the patient about what happened without speculating on whether anyone is to blame
- tell the patient that a full review of the event will occur to determine what happened and to prevent this type of adverse event from happening again
- provide regular feedback about the process of the event review and its results

In the United Kingdom, the NHS guidelines include the fact that there is exemption from disciplinary action for professionals who report adverse events or medical errors, except where there is a criminal offence.

The key concept is that open disclosure is about prompt, compassionate, timely, and honest communication with the patient and family. It is not a one-off event. It is a two-way, ongoing process.

The initial goal of open disclosure was to reduce litigation from adverse events. This has been effective, but a more important goal is to learn from adverse events so that the system can be improved to help reduce errors. While open disclosure is important for the patient, it also provides opportunities to improve the system in which we work so that other patients will be safer.

Current State of Open Disclosure

Genuine Regret

Have you ever called a company to complain about their product, have the person you called say, "I'm sincerely sorry about that," and then promise to do something to resolve your issue, but nothing happens, and you hear nothing further? It makes you wonder if they are reading their "apology" from a script of answers to customers provided on their computer screen. There is no acknowledgement that a mistake has been made, and the apology is not sincere.

We have to decide whether disclosure of our errors is a self-protective, risk-management response that doesn't do anything to help the patient—or a disclosure that is framed within a morality of authenticity. We need to be aware of the difference between an inauthentic response to medical harm and a sincere apology. We also need to remember that the person who has been harmed has the freedom to choose between acceptance or rejection of our apology. The authentic sincerity of our apology will go a long way toward it being accepted. On those occasions when acknowledgement of our error and our apology may not be accepted, it should make no difference to the level of care and the amount of communication we provide. And we should still make the same efforts to solve the underlying problem that led to the error.

Forgiveness and retribution are concepts that are sometimes thought to be incompatible. If we have been harmed, we may be able to forgive the person who harmed us, but we still want the system fixed, perhaps to be compensated, and for the person who caused the harm to be disciplined if the harm was attributed to reckless care.

So, if we are going to admit our mistake and apologize to the patient, we need to learn to do it properly. A shoddy apology or one given for the wrong reasons can do more harm than good.

Barriers

Although there is wide appreciation of the need for and the value of open disclosure, there are a number of barriers that can prevent it from happening such as:

- fear of litigation
- facing disciplinary action as a result of our open disclosure
- senior management not having a clear commitment to open disclosure and not promoting it as a core policy
- no education about how to talk to patients about an error
- lack of support for staff when they do open disclosure
- fear of risk to our reputations
- losing respect from peers and colleagues
- a legal environment that may fuel fear of litigation if we make an error and admit it
- personal emotional distress

As individuals, we may fear that honesty could expose us to the risk of discipline, particularly if we work in a toxic environment where a culture of blame predominates. We may fear litigation, loss of respect from colleagues, and loss of referrals. Acknowledging our error may make us feel vulnerable and less confident.

We may also feel that there is lack of training and a lack of management support in our organization to undertake an open disclosure process. We may work in an organization that still has a culture of self-protection and secrecy and where the patient is not the number one priority.

Who Should Tell the Patient?

We know that most errors involve a sequence of events in the care system with the person who actually caused the error being the last person in that sequence. It is clear that this person should be involved in the open-disclosure process. However, if it is a junior person, perhaps with no training in open disclosure, he/she should be supported by a more senior person who can be part of the process and then provide constructive feedback. It's a balance between not overwhelming the patient and family with a complete team and ensuring that junior staff are supported in the process by clinical staff who are experienced in this area, have good communication skills, and model honesty, transparency, and empathy.

It's also important to be aware that open disclosure is not a one-off event. It is an ongoing process that serves to support both the acute and long-term needs of the patient and/or family.

The essential steps in acknowledging an error and apologizing involve:

- Telling the patient what happened. We can leave the details of *how* and *why* until after careful analysis of the event has occurred. Many times, that first conversation is sharing that we don't know what has happened but that we are taking this adverse event very seriously and promise that we will find out what happened. Open disclosure is an ongoing process. At this stage, it is important to only give the known facts, avoid speculation, and tell the patient that this is an ongoing process and that you will be in touch regularly as more becomes known.

- Take responsibility. If you are a more senior person in charge of the patient's care, you should assume responsibility even if you did not actually make the mistake that caused the harm.
- Apologize if appropriate. This is an essential part of taking responsibility.
- Explain what will be done to help this patient.
- Explain what will be done to prevent future events like this so that other patients will be safer.

We Also Suffer When Errors Occur

Providers never leave for work in the morning intending to cause harm, so when it occurs, we feel devastated. There may be a sickening realization that we've made a mistake that has harmed a patient, which is quickly followed by anxiety about what to do, who to tell, how to tell, when to tell, or even how to cover up our error.

If it's a serious error, we may lose confidence. We may fear that we will be punished or that our colleagues and seniors will trust us less. Sadly, our colleagues may not support us. Reactions by peers following an error by a colleague can include criticism, gossip, lack of respect, and reluctance to trust us in other clinical situations. Because we have high standards, there is a danger that we may cope in dysfunctional ways, such as being angry at ourselves, projecting the blame onto others (or even the patient), becoming defensive, losing confidence, burning out, or seeking consolation in alcohol or drugs. While the patient is the main victim, the clinician can also be injured and become the "second patient" or second victim.

We need to support our colleagues when they make a mistake. We shouldn't minimize the seriousness of their error, but we should be prepared to listen to them, to be empathic and supportive, to encourage open disclosure, and to avoid the temptation to gossip or undermine them.

Lessons Learned

The concept of health professionals admitting their errors and using mistakes as an opportunity to fix the system is relatively new, but it has a lot going for it. Making care safer has a direct effect on reducing future malpractice claims. There are many examples where it has reduced malpractice claims and litigation and actually saved the health system money. More importantly, it promotes transparency and honesty in all aspects of patient care. However, the health care systems we work in still have a long way to go before the patient's best interests are at the center of every aspect of care.

Rick van Pelt is an anesthetist whose patient, Linda, had an unexpected reaction to the anesthetic followed by a cardiac arrest during an orthopedic procedure on her ankle. Cardiopulmonary resuscitation was not effective, so she was immediately placed on heart-lung bypass until her heart function was restored. Instead of waking in the recovery ward with a bandaged ankle, Linda woke in intensive care with a wound where her chest had been opened to institute bypass. She did not receive an adequate explanation for what had happened to her.

Rick received no support. He was blocked by his supervisors from seeing Linda while she was in the hospital and told to resume his normal duties as if nothing had happened. Eventually, Rick sent a letter to Linda's home saying he was sorry. When they met, Rick apologized—and Linda offered forgiveness—which resulted in healing for Linda and for Rick. They have since become passionate advocates for open disclosure.

It didn't have to happen that way. Once hospital administrators and health professionals recognize the importance of being honest when things go wrong, patients are treated with honesty and respect. Staff are supported. Systems are changed to make it safer for other patients.

This next generation of health care leaders, the first generation that has been exposed from early training to concepts of patient safety and improving the quality of care, is the generation that will make the difference.

The next generation of clinical leaders is likely to be different from past generations. The old-style clinical leaders are excellent clinicians, intelligent, astute, and respected. They are proud of their clinical autonomy, comfortable with their decisions, and often not responsible to anyone but themselves for their work. They lead their teams fearlessly, shoulder much of the responsibility themselves, and have a strong influence on their junior staff (so much that junior staff may unconsciously adopt their characteristics, the good ones and the not-so-good ones). They leave administrative things to the administrators (for whom they may not have a great deal of respect) and don't worry much about costs, proudly declaring that "my patients' needs always come first" although, in reality, the patients' needs sometimes have to fit in around the clinicians' busy schedule and other priorities. Of course, this is a parody, but we do recognize parts of these features in people who lead, perhaps even in ourselves.

This is because, traditionally, patient care has been based around what doctors do instead of than around what patients need. In contrast, the new generation of clinical leaders will ensure that the group they lead will, first and foremost, want to make a difference for good in the lives for whom they care. They will work differently. They will still have a strong emphasis on the highest possible level of clinical care, but they will also focus on delivering care efficiently and reducing waste. Delivering care efficiently means better coordination of care to avoid duplication, reducing administratively complexity, and communicating effectively at all levels. Reducing waste includes not over-investigating by doing investigations that do not help the patient and avoiding treatment that does not help the patient or is unnecessary. And it includes avoiding error, reducing harm, and being open and transparent.

The new generation of clinical leaders will:

1. Lead reform by putting the patient first so that clinical decisions will involve the patient and will always have the patient's best interests at heart. They will see services from the patient's point of view.
2. Realize that even networks become silos in their own right and will build bridges across networks because this will benefit understanding and lead to better care.

3. Create a culture of safety where errors will be seen as opportunities for improvement and where open disclosure is supported and is part of the accepted process of care.
4. Be a mentor within the team, be a motivator of others in the team, and be a facilitator.
5. Be a good communicator to patients, to the team, and with administration. They will know that stories are a powerful form of communication since stories can influence hearts as well as minds, and they will remember that being a good communicator also means being a good listener.
6. Be a good clinician, be aware of relevant research developments, and be able to use and interpret data to improve patient care, improve efficiency, and reduce waste.
7. Take overall responsibility for managing finances, even if this is delegated to a professional with more expertise.
8. Be honest, open, and ethical, displaying integrity in all dealings.
9. Be able to learn from experience.
10. Emphasize and model to the team that it is the patient, rather than the clinician, who is at the center of the stage, reminding the team that "it's all about the patient."
11. Know how to be constructively subversive to create change— how to rock the boat without tipping it over.

Of course, many of these characteristics are found in the current and earlier generations of clinical leaders. My hope is that the next generation will embrace and adopt them because they will know that this produces better patient care and can even reduce costs.

Three Short, Concluding Messages

- Confessions are hard. They raise difficult feeling in us as well as in those who have been harmed. Open disclosure is not an option when we cause harm. It is an integral part of good care. Open disclosure is about taking responsibility. It is about prompt, honest, compassionate communication. It is about seeing an error as an opportunity to improve the system.

- A new generation of clinical leaders is coming. Quite a few of them are already here. You don't have to be young to be part of this new generation—you just need to look at things differently.
- It's all about the patient, always, every time.

CHAPTER 8

Can a Conversation Change an Outcome?
Tim McDonald

Initial Reflections

An Informed Consent Conundrum

In 2011, an anesthesiologist approached a patient in the Anesthesia Pre Evaluation Clinic to discuss the patient's upcoming thyroid surgery and associated anesthetic options. Two weeks earlier, the surgeon performed a needle biopsy that showed the presence of a cancerous nodule that would require surgical removal as the first step toward treatment of the cancer.

Of note, the surgeon's documentation indicated that he intended to perform the surgery using a minimally invasive approach that would require the use of the da Vinci surgical system. As the anesthesiologist reviewed the records, he became acutely aware of a couple of important issues. First, the use of the robot would cause the anesthesiologist to be separated far from the airway of the patient during surgery because of the placement of the arms of the robotic apparatus. This is an important consideration in head and neck surgery because the anesthesiologist prefers to be as close to the patient's airway as possible in the event of common airway complications during the surgery.

For this particular surgery, the anesthesiologist possessed some other important knowledge that posed a second important issue and problem for consideration. This patient was the first patient upon whom the surgeon would use the robotic device for thyroid surgery—and no publications existed at the time of this surgery to validate its effectiveness or superiority to a standard non-robotic approach to this type of surgery. This approach did offer the patient two smaller non-midline incisions versus the traditional midline two-inch incision in the middle of the neck. According to the surgeon's note, the patient indicated a desire to avoid a midline incision for cosmetic reasons.

Both of these issues were swirling in the anesthesiologist's mind as he entered the patient's room with the anesthesia preoperative history and physical and informed consent forms. Would he raise these issues during his visit with the patient? What would an alumnus of TTE do?

TTE Scholar Reflections about Informed Consent

Following the viewing and discussion of the powerful patient safety video "From Tears to Transparency: The Story of Michael Skolnik," TTE learners provided some poignant reflections about the video and their own personal experiences related to informed consent and shared decision-making prior to coming to TTE. Their reflections provide some insight into the current state of medical education and practices around the informed consent process.

As Alana Gilman (2014) aptly points out in the blog post, "Everyone is Special to Someone," clinicians often forget that for patients who just discovered they have cancer and now need a surgical procedure, "this may be the worst day of their life" and we clinicians often forget this may be true "for each patient we encounter."

One TTE faculty member, Dr. Roger Leonard, made a powerful impression on this student around this "forgetfulness of the patient's perspective" as the student recalled Dr. Leonard "profoundly mentioning that the informed consent is a red-flag moment that deserves full attention, just like critical lab values of critical vital signs."

Gilman astutely said, "This seemed like a novel point of view compared with the treatment of the topic up until this point of [our] education and the little training [we've received] surrounding it." Importantly, Gilman also picked up one of the most important lessons from the Michael Skolnik video:

> The informed consent is a monumental event, but as I learned yesterday in the film, even more than an event, a process. It is an ever-evolving conversation that is grounded in a multidirectional communication between all interested parties, including the traditional players such as the physician and the physician and, oh yeah, sometimes the patient but mostly the physician. No! This is a conversation with the patient, and anyone and everyone who is there supporting that person through this difficult time and may have a clearer head while making potential life-altering decisions with unsure outcomes. It also involves the physician but should welcome the input and assistance of other members of the health care team, especially the bedside nurses, who are there with the patient hour after hour, long after the initial informed consent conversation has ended, when the patient is finally able to think clearly enough to have questions.

Concluding her reflection, Gilman goes on to describe the informed consent process as a "beautiful opportunity to sit down with a patient and guide and advise them through a challenging decision for which they commonly have little expertise." It seems Gilman is now well prepared for situations as the one described about our anesthesiologist and his ethical conundrum.

Over the many years of helping to host The Telluride Experience, I recall many scholars admitting that the informed consent process had been lost on them. Consent was a task and not a process, in their minds, and this one page "fill-in-the-blank" piece of paper just needed to get signed. Many would sadly confess they felt a bit guilty discussing the risks, benefits, and alternatives of an invasive procedure when they, themselves, did *not* know the risks, benefits, and alternatives. These forms were perceived as legal documents intended to protect the physician while carrying little value for the patient or family.

Fortunately, in the midst of TTE, most learners come to a wonderful new understanding of informed consent, especially after the showing of the Michael Skolnik film and the passionate follow-up discussions. The concept of consent becomes better understood as the open, honest, and transparent dialogue between the provider and patient with their loved ones. The conversation—where there can be assessment of the patient understanding of their illness and treatment options—takes place at the patient and family level. Informed consent is an opportunity many of the scholars would concede to discuss and explore options and, hopefully, arrive at the treatment pathway that best meets the patient's needs. They learn that informed consent is not something you just go get!

These learners demonstrate, in these discussions, the power of changing mind-sets. Like Gilman, these learners also seem well prepared when they leave TTE to tackle the difficult consent issues as those described above.

Another learner reflection comes from "Good Communication Can Save Lives," a blog post by Mihai Dumitrescu (2014) that describes the mental model of many clinicians who practice physician-centric instead of patient-centric medicine. These words seem to reflect the thoughts of those who do not put the patient in the center of all they do. "You're so happy to have the opportunity of doing a procedure, so nervous and often uncomfortable that you forget you're dealing with people's lives." Dumitrescu goes on to ask, "How many times were you so excited about the procedure you were about to perform that you forgot to ask the patient (or parent in a minor's case) how they feel?" These words seem to reflect those of the surgeon who is anxious to do their first robotic-assisted invasive thyroid procedure.

With a thoughtful expression of self-awareness, the state of the current informed consent process is described from this resident's perspective:

> We are so busy as residents and often, sometimes too
> often, we forget about the patient. We forget about

the family, and we forget to slow down, and explain to our patient the *what, why, if's and how* the procedure, treatment, or intervention around the corner is going to happened ... It is almost like not including them in the decision process. We only need their names, their signatures, so we can move on with the procedure and our daily activities.

As with the other learners, "The Story of Michael Skolnik" and the follow-up discussion had a profound effect. Dumitrescu says:

Today we opened our eyes to see the consequences of poor communication, the effects of not slowing down, and how the lack of experience [impacts] someone's life. Further, when you hear a story like Michael's, your heart stops beating for a moment, your eyes are filled in tears, and the only thing you can think of is "this can't be true" ... But, yes, it's true! Unfortunately, medical errors are killing too many people, and communication is one of the main causes ... We need to work together; we need to communicate better; we need to find solutions; we need to advocate for patients and families ... We need to take the role of guardians, implement a culture of safety and zero harm!

In the context of these reflections, it's time to return to the case above.

Outcome of the Informed-Consent Conundrum

The anesthesiologist, inspired in part by the teachings contained within "The Story of Michael Skolnik," entered the exam room of the patient in the Anesthesia Pre Evaluation Clinic. He explained to the patient that the technique the surgeon recommended would add significant risk of a "difficult-to-manage" airway complication.

During the course of the conversation, the patient asked whether the anesthesiologist had ever been involved with a procedure performed in the manner recommended by the surgeon. He was also asked whether the surgeon had performed such a procedure previously. The anesthesiologist pondered those questions and decided to answer openly and honestly and explained that neither he nor the surgeon had ever conducted a thyroid surgery in the manner that had been recommended. He did share with the patient that the surgeon was otherwise excellent at thyroid surgery and, to his knowledge, had a very low complication rate when performed in the standard fashion.

The anesthesiologist went further and asked the patient about her preferences and the degree to which the cosmetic outcome was important to her. He confided that robotic surgical procedures, especially when performed by a novice on the robot, often take three times as long as the standard approach, but the associated incisions are a bit smaller. Importantly, the anesthesiologist shared this conversation with the surgeon.

Armed with this information, the patient refused the approach the surgeon had recommended and, instead, agreed to undergo the surgery with the standard, traditional technique. She did not become the surgeon's first robotic thyroid surgery patient. The surgery proceeded without complication, and the patient experienced an excellent recovery. Of note, however, the surgeon requested that the anesthesiologist who was open and honest with his patient never provide anesthesia for one of his patients again.

More progress is needed to change the culture in medicine, and this case emphasizes the need for programs that "educate the young" and give them the tools to practice transparent patient-and-family-centered medicine. Based upon the reflections of the TTE scholars, it seems they, too, will be champions for the needed culture change and, if confronted with similar situations, will take the time to listen to a patient in similar circumstances and have the courage to speak up and be open and honest. They will engage in a conversation.

Going Forward

Knowledge is power, and going forward, it is important that the future leaders in patient safety possess a deep knowledge of the concepts of informed consent. The use of patient stories, such as Michael Skolnik's, is critical for reinforcing the importance of these concepts and the take-home messages and learning to focus on the consequences for patients and families when they choose certain treatment paths.

Also importantly, from the didactic perspective, the following key points about informed consent are critical for the learner to understand.

What Is Informed Consent?

Informed consent is an information-sharing process that is based on the moral and legal premise of patient autonomy—the patient has the right to make informed decisions about their own health and medical conditions. Voluntary informed consent is necessary for treatment and for most medical tests and procedures. Failing to obtain informed consent before performing a test or procedure on a patient is considered battery—the unlawful touching of another person. Penalties for battery can be both civil and criminal, depending upon the egregiousness of the touching without consent.

For many types of interactions (for example, a physical exam with your doctor), implied consent is assumed. This is true in many facets of life. When someone chooses to play a game of basketball in the neighborhood, it is assumed that physical contact will occur and that those who choose to play are consenting to that contact. Assumed consent is also present when patients present with emergency situations where a procedure or treatment is needed to prevent serious harm. For more invasive tests or for those tests or treatments that are associated with more than minimal risk, a written consent is expected whereby documents will be signed by the person who will undergo the procedure.

Elements of Informed Consent

There are several components of an appropriately obtained informed consent. Informed consent is a process and not the piece of paper upon which the process is documented or memorialized. The process of informed consent should begin with adequate time for the patient and family to consider the consequences of their decision. Many consider it inappropriate to begin this process on the morning of a surgical procedure. The person providing consent must have the capacity (or ability) to make the decision.

The medical provider must disclose information on the treatment, test, or procedure, including the expected benefits and material risks, and the likelihood (or probability) that the benefits and risks will occur. This also includes the risk of not undergoing the procedure or test. Material risks are those that a reasonable person would consider pertinent in making the decision. These include frequent, albeit nonserious, complications and less frequent but more serious complications. The consent must be voluntarily granted, without coercion or duress, and can be revoked at any time—even moments before an invasive procedure. The person providing consent must have the opportunity to ask questions and have those questions answered truthfully.

Decision-Making Capacity

The patient giving their consent must have decision-making capacity. Some of the components of decision-making capacity include:

- the ability to understand the treatment options (the information must be conveyed in a way that considers the level of health literacy and the preferred language of the patient)
- the ability to understand the consequences of choosing each of the options
- the ability to evaluate the personal cost and benefit of each of the consequences and relate them to their own set of values and priorities (also must be of legal age to provide informed consent)

If the patient is not able to engage in all the components, family members, court-appointed guardians, or others (as determined by state law) may act as "surrogate decision-makers" and make decisions for the patient.

To have decision-making capacity does not mean that the patient will always make "good" decisions, or decisions that comport with the recommended treatment. Likewise, making a "bad" decision does not mean that the patient is "incompetent" or does not have decision-making capacity. Capacity, or competency, means the patient can understand and explain the options, their implications, and give a rational reason why they would decide on a particular option instead of the others.

Information Sharing

The doctor or health care provider must share enough information so that the patient can make an informed decision. The information if shared would be enough that a reasonable person could make an intelligent decision and should include the risks and likelihood of each of the "material" risks, benefits, and likelihood of benefits. Questions should be fully explained, in language and terminology that the patient can understand.

Informed Consent: The Right to Refuse Treatment

Except for legally authorized (court-approved) involuntary treatment, patients who are legally competent to make medical decisions and who are judged by health care providers to have decision-making capacity have the legal and moral right to refuse any or all treatment. This is true even if the patient chooses to make a "bad decision" that may result in serious disability or even death.

In Summary

The anesthesiologist, in this case, followed the principles and met the elements of an appropriate patient-centric informed consent process. He shared information that reasonable patients would want to know— namely, the added risks related to the airway, the extended length of the procedure, and the fact that this would have been his first case and the surgeon's first case of using the robot to perform this surgery.

He also listened to the patient and came to a consensus based upon her needs and preferences and not his own. Finally, the anesthesiologist had the courage to share this information, knowing that such sharing might strain his relationship with the surgeon.

For the learner, it is important to compare and contrast the actions of the anesthesiologist in this case with those of the surgeon in the Michael Skolnik case and reflect further on how many of the elements of informed consent were not met and how the approach to the surgical procedures were provider-centric and not patient-and-family-centric. It is important to recognize, in Michael's case, that a robust patient-centric conversation would have saved a life.

CHAPTER 9

Training the Next Generation

Anne Gunderson

Blog Posts Written by Health Care Learners

Let's Put the Numbers on the Board
Samantha Nino
June 17, 2014

Too many times during this roundtable so far, have I heard a fellow student exasperate, "Why aren't we learning this in school?" The knowledge we're acquiring is invaluable, and we are collectively shocked and disgruntled that these topics are not effectively breached, especially during the didactic half of the medical school curriculum.

The Way We've Always Done It
Christine Beeson
June 17, 2014

John Nance stated, "The most dangerous phrase in medicine is 'this is the way we've always done it.'" He went on to say medicine is not solely a profession. Instead, it is "a calling wrapped around a profession with a purpose of service to humanity." How eloquently and beautifully worded!

Instead of cowering away from the charge we've all been given to tackle this mountain (pardon the pun) of a problem, we should enter our callings head-on with power and confidence.

For me, this means being a leader as a physician—but not in the way physicians are traditionally thought of as leaders. No longer is there a culture of "what the doctor says goes."

In many ways, we are still trained to memorize *all* the drugs, *all* the adverse effects, and *all* the details of *all* the disease processes … but why? Why don't we change the very core of medical education to embrace the technological advances we have made in nearly every other industry? Why don't we train our medical professionals in a way that utilizes technology to empower the health care provider and not simply provide a means to keep medical records? We must change our systems to incorporate teamwork, encourage the utilization of technology, and reframe the entire medical hierarchy.

Culture Change? Which One?
Garrett Coyan
June 21, 2013

In order to change the safety culture of the hospitals and health systems we work in, we need to create providers who are trained, knowledgeable, and willing to implement the changes needed to provide quality, compassionate, and safe care to our patients. It is for this reason that I will be speaking with my dean, director of quality improvement, all my mentors, and as many of my classmates as will listen about making this information a mandatory and testable portion of our curriculum. It is only then that we can hope for young medical students, residents, and physicians to be competent and comfortable enough to speak up when medical errors are made and confront them head-on and honestly with our number one partner in health care: our patient. Culture change will not occur until we start demanding it!

Roger That
Evan Skinner
July 31, 2014

Closure of communication loops between nurses and physicians is probably one of the strongest initial interventions we could foster together in order to bridge some of the distance between our professions.

Proud, Humbled, and Somewhere in Between
Christine Beeson
June 16, 2014

To say today was educational, informative, and life-changing would be an understatement. We have medical students, both MD and DO, from multiple schools. We have nursing students from all corners of that vast field (NP, PhD, master's, BSN). We have master's in health administration students. We have students who are working on MD/PhDs or MD/MPHs. We have students who have done years of research on patient safety and quality improvement. We have students who have worked with the underserved both in our country and overseas. We also have students from other nations.

At times, I catch myself wondering how I was chosen to attend this conference—I can't compete with those people! My list of experience is short, to say the least! But in response to those thoughts, I remember how humbled and grateful I am for being given this opportunity. The sharing of ideas going on in each discussion is truly incredible! The combination of differing opinions, expertise, and various backgrounds and experiences in patient safety make every one of our key topics thought-provoking, inspirational, frustrating, overwhelming, and promising.

Initial Reflections

In a successful quality and patient safety health science curriculum, it is important to understand and recognize interdisciplinary teamwork skills. While many medicine and nursing deans agree they should provide interdisciplinary clinical learning experiences, they acknowledge that they have no idea how to go about making the idea a reality. Unfortunately, the deans often settle with creating a six-to-eight-hour "safety day" to supplement individual IHI modules or other "do-it-yourself learning" Or worse, running one simulation in the name of "safety" and checking the box as "done." Neither of these educational choices begins to meet the needs of the learners. Granted, the exercises might get the institution through accreditation requirements, but it is ineffective for measurable learning. The educational environment remains entrenched in recognizing and rewarding individual accomplishment over the power of teams.

Historical Background

Changing the health care culture to one of quality care and patient safety would require changing the culture of health science education. It would not be an easy endeavor and would require considerable energy, education, and resources. The aviation industry, however, recognizes team mechanics and competencies as important components in a safety culture and sends flight teams to work and train together. If they can do it, why can't health care? To prepare future health care learners to function effectively as members of a care team, the faculty and deans must step away from the past and look deeply into the current state of heath care. Medical errors are now the third leading cause of death and account for more than four hundred thousand deaths per year (James, 2013) and as many as one-third of hospitalized patients may experience harm or an adverse event, often from preventable errors. I would suggest the data speaks for itself.

Current State

The United States continues to have health care quality problems despite all its spending (Becher & Chassin, 2001). Escalating medical errors in the United States affect who that accesses the health care system. Since the release of the Institute of Medicine's "To Error Is Human" report (IOM, 1999), discussion has been about modifying the health science education system to address ongoing patient safety concerns. Improving health care practices begins with improving health science education.

The American Association of Colleges of Nursing (AACN) and the IOM have described the health care delivery system as "broken." The IOM called for a change in educating and training physicians and to address the problems associated with quality, access to care, and patient outcomes in the health care system. Competencies for optimal patient care outcomes in the clinical environment include knowledge, skills, and attitudes in critical disciplines not traditionally taught in medical schools (Cooke et al., 2006). Unfortunately, reforming health care education to address safety and quality issues presents a challenge to educators.

The United States has entered an era in health care where understanding the physiology and treatment of disease is not enough. There has been discussion in both the public and the private sectors regarding ways to modify the health science education system to address growing concerns. Although patient safety has been increasingly recognized in quality care, systematic safety education for health care professionals is lacking (Gould et al., 2004). Collaboration and communication training between health science students remain unaddressed. Wachter (2004) noted, "The shortcomings that must be addressed are deeply entrenched in the tradition and culture of the institutions and organizations that compose the medical education system."

Medicine, nursing, and general health care programs have long been viewed as a science, but they generally do not train together. Curricular and goal-oriented learning programs have traditionally utilized instructional modalities such as lectures, modeling, supervised practice, mentoring, and experiential learning to augment individual study. In this new world, we must adopt the new skills and the attitudes necessary to meet the changing needs of patients in a medical environment. Health science learners could utilize a new educational program to understand, recognize, demonstrate proficiency in, and assume a leadership role in patient safety and quality outcomes initiatives.

Changing the science culture to one of quality care and patient safety would require the introduction of team-building values and skills to the learners. This is a new concept, especially for medical students and resident learners. The learning experience can offer opportunities for medical learners from multiple disciplines to work together in simulated or standardized patient scenarios that allow them to practice and share feelings about effective team-building skills. The health care environment requires a systems-oriented, multidisciplinary, team-driven, patient-centered approach for optimal patient success. If medical educators are going to change the health care culture into a culture of optimal quality and safety, it is important that health science learners begin to understand and recognize the direct link between scientific advances and improving health care outcomes.

Patients utilize teaching hospitals for surgeries, emergent care, and complex treatments. Teaching hospitals are where medical knowledge continuously evolves and new cures and treatments are found. These hospitals are the training ground for more than one hundred thousand new physicians and other health professionals each year. Teaching hospitals, also frequently referred to as academic medical centers, are the clinical side of the higher education system that encompasses medical schools in the United States.

"The purpose of medical education is to transmit the knowledge, impart the skills, and inculcate the values of the profession in an appropriately balanced and integrated manner" (Cooke et al., 2006). Medical educators should find ways to align medical school curricula with the ongoing quality care and patient safety initiatives to help accelerate the cultural change. Few medical schools have attempted to give their students the skills they will need to measure the outcomes of care they will provide to their patients. Quality measurement and improvement, cost-effectiveness evaluation, and the measurement of patient satisfaction are all components of contemporary health care systems designed to deliver high-quality, user-friendly, affordable health care.

Lessons Learned

1. Safety education for health care professionals is lacking.
2. The health care delivery system is broken.
3. True patient care occurs when the patient is the center of everything.
4. No longer is there a culture of "what the doctor says goes."
5. All providers of care must train together.
6. Improving health care practices begins with improving health science education.

CHAPTER 10

Neglecting Our Caregivers

Roger Leonard and Sherri Loeb

Blog Posts Written by Health Care Learners

Where Is Our Support?
Nicholas Clark
June 12, 2014

On Wednesday, we discussed the case of "Sally," a nine-year-old girl who died because of medical errors. Regardless of how you look at it, this is a tragedy. In our discussion, the presenter described why this resident was "set up to fail." The resident had undergone numerous emotional battles in the prior months on the wards and in the ICU, had struggles outside of the hospital, and ultimately quit the residency program as a result of Sally's death, but there was not one discussion on Wednesday about how we should care for our residents.

Unfortunately, this resident's story is all too common. Many of our TTE attendees sympathized with the resident, outlining how similar their experiences have been to Sally's resident. I too can look back and see myself in that position. It is well-documented in the literature that residents, regardless of profession, develop higher rates of depression and suicide than the general population as a result of our profession. Up to

20 percent of residents and medical students will face depression, and up to 74 percent of residents will face burnout. Those residents who battle with depression are six times more likely to cause medical errors than those residents who do not suffer with depression. While it is easy to point the finger at the resident or the system for causing medical errors, and, at the same time, provide support for the family and the patient, there are few programs in place that support our residents and medical students. These are individuals who choose the medical field to cure pain and suffering—not cause them. When residents discover that they have harmed someone, despite their best knowledge, skill, and intention, it is absolutely devastating!

Do not get me wrong. I completely agree that we should first tend to the patient and family affected. However, we cannot forget about the second victim who was harmed: the resident. We need systems that automatically fire to debrief residents when harm occurs so they can learn from the event. We need systems that automatically fire to find systematic solutions to the problem so no other patient is harmed. We need systems that automatically fire to provide support for patients and families who suffer harm. However, we also need systems that automatically fire to provide emotional support for our residents when they are involved in a case of patient harm. Finding that you were part of the cause for patient harm can have a devastating and lasting effect. Residents are already set up for depression, which is a setup for further medical errors. These two perpetuate each other in a never-ending cycle. We need to break the cycle. We need support!

Communication, Empathy, and Compassion
Rupa Prasad
June 10, 2015

Reflecting on today's sessions, the idea of empathy and compassion as integral components of effective communication and patient safety has me thinking about the current culture of health care. There is no doubt that empathy and compassion are of the utmost importance in health care and that physicians and nurses as individuals who enter this field value these attributes as such and aim to embody them. Yet, how is that so often we as health care providers seem to forget to put these concepts into action in our daily interactions with patients?

This is not to say that we intentionally neglect patients' feelings and concerns. I believe that most truly care for their patients and intend to be empathetic and compassionate, but going through my third year of medical school has changed my perspective on if we're succeeding in this mission.

Before starting our third year, our faculty discussed with us the concept of burnout and the trend toward a decline in empathy as we progress through our medical education and training. To be honest, at the time I thought, *I understand what you're saying, but that won't be me.* Now, looking back on my third year, I can see how the point they were making is sadly very true.

As I went through my clinical rotations, I saw firsthand the reality of burnout and its effects on empathy, compassion, and patient care as it affected the interns, residents, and attendings around me–and then, by the end of the year, on myself as well. There exists a culture in health care that is complacent with a decline in empathy. It's frightening to realize this is the case and that I am not immune to it.

This culture and trend cannot be good for patient safety since empathy and compassion are important factors in effective communication with patients as well as our colleagues, and without effective communication, we cannot keep our patients safe. I don't have the answers to this problem now, but being here in TTE has me thinking about how we can begin to solve it and will definitely have an effect on how I approach situations where I see that empathy and compassion are lacking in the future.

Initial Reflections

Health care professionals are expected to serve patients in stressful situations, while being short staffed, overworked with long days, and often with less than enough sleep. When something bad happens, be it error or just normal disease course, we are expected to immediately "get back on the horse" like nothing ever happened. We are taught to have a stiff upper lip that only contributes to the feeling of isolation and lack of support. As these young professionals wrote in their blog posts, this causes residents and students to question why they went into the profession.

With altruistic intentions, empathy, and a passion to make a difference in the lives of others, we are learning to suppress our emotions in a diminished culture that tolerates insensitivity and devalues our humanity. Health care leaders are aware. Unfortunately, a dysfunctional reimbursement system causes too many leaders to become managers who focus more on the financial profit and loss vis-à-vis the human profit and loss.

Police and fire services are examples of professions that provide immediate support when something bad happens. A fire department or police department chaplain gives support and counseling for job-, personal-, and family-related problems. After a significant tragedy, the firefighter or law enforcement officer can be removed from the direct line of work for needed counseling. In medicine, you may be caring for a patient for weeks, and when you lose them, you are expected to immediately turn around and act as if nothing happened.

> The fault, dear Brutus, is not in our stars, but
> in ourselves, that we are underlings.
> —Julius Caesar 1, ii, 140–141

The quote by Cassius in Shakespeare's *Julius Caesar* may have a couple of interpretations with regard to burnout, harm, safety, and quality of health care today. At first take, it suggests that individual accountability is required to solve the problem rather than changing "the system." This assumption is wrong. Our intent is the opposite. Health care professionals, individually and collectively, have the power to change the health care system that creates a new destiny. Patients and families are powerful allies in this endeavor, and we are ethically accountable to them. Together, we can change a medical culture that devalues and tolerates the loss of compassion, empathy, and mutual caring. Care for the caregiver is essential, whether for a family member, medical student, resident, attending, nurse, or allied health professional. We need to change the health system's conditions or else we will continue a 20 percent incidence of depression and 74 percent incidence of burnout with devastating, preventable harm and waste.

Historical Background

> Physician, heal thyself.
> —Luke 4:23

The ability of a physician to provide optimal, ethical care requires the virtues of integrity, compassion, self-effacement, and self-sacrifice. Like any group or profession, medicine is composed of individuals who span the emotional and behavioral spectrum. Emotional intelligence is an attribute that varies widely among health professionals. With whatever nature-nurture trophies or baggage they bring, physicians-to-be begin a long, stressful period of education and training that often exceeds a decade. It is typical to believe that all previous generations of physicians survived such an arduous challenge.

The truth is not so. The incidence of depression for medical students (Brown et al., 1986) and resident physicians (Shanafelt et al, 2002) is higher than for the general population. The relative risk of suicide for male physicians is roughly twice that of the general population, and the relative risk for female physicians is roughly four times that of the general population (Lindeman et al., 1996). Among medical students who screened positive for depression, less than one-quarter sought help. Among students with suicidal ideation, fewer than one-half received treatment (Givens, 2002).

In the following paragraphs, we will take a more recent look at depression and burnout in medical students, residents, practicing physicians, nurses, and families respectively. Several factors influence the variability of the data such as the sample pool and response rates. Importantly, the tools to measure depression differ across studies. However, there is uniformity in using the Maslach Burnout Inventory (MBI). The MBI consists of twenty-two questions grouped to assess emotional exhaustion, depersonalization, and personal accomplishment. For reliability and consistency, most authors combine the emotional exhaustion and depersonalization scores to determine burnout, using personal accomplishment as a sidebar.

Medical Students

In a study of 1,701 medical students from five geographically diverse medical schools in the United States, researchers sought to determine student well-being and determine if there were differences due to race or ethnicity (Dyerbye et al, 2007). Nearly half of the students experienced burnout, and an equal number reported depressive symptoms. There was no difference in the incidence of depressive symptoms comparing minority versus nonminority students, but minority students experienced a significantly higher incidence of burnout.

Resident Physicians

A literature review of resident physicians' stress consistently finds that the majority of young physicians experience burnout at some point in their training (Thomas, 2004). Multiple studies observe burnout in approximately 75 percent of pediatric residents, 65 percent of internal medicine residents, 56 percent of orthopedic residents, and 47 percent of anesthesiology residents. Pediatric residents report a 20 percent incidence of depression. Further, depressed residents were six times more likely to make a medication error with the potential to cause serious harm in approximately half of the cases (Fahrenkopf, 2008).

Practicing Physicians

In a survey of more than twenty-five thousand practicing physicians, researchers found that nearly half experienced burnout, and more than a third had a positive screen for depression (Shanafelt et al., 2012). The highest rates of burnout occurred in emergency medicine (65 percent), internal medicine (54 percent), neurology (52 percent), and family medicine (51 percent). There is concern even for those specialties with the lowest burnout rates: dermatology (32 percent), pediatrics (35 percent), pathology (37 percent), and radiation oncology (38 percent). Compared to the general adult American population, physicians were significantly more likely to experience burnout.

Nurses

Within the past decade, the twelve-hour shift has become a popular option in nursing, believing that it provides greater continuity of care, greater lifestyle flexibility, and reduced costs. The evidence to support such benefits is inconclusive.

According to Dawn Kettinger, a spokesperson for the Michigan Nurses Association, burnout can be caused by several factors, including lack of social support, inability to control one's work schedule or assignments, a chaotic or monotonous job, and work-life imbalance. Kettinger believes that short staffing is often the primary contributor. As hospitals try to cut costs, that tactic backfires and becomes a patient safety issue.

Nursing burnout is also affecting higher rates of turnover in the profession, ultimately causing a shortage of qualified professionals while future demand grows. Short staffing caused by increasing burnout rates results in higher infection rates, higher preventable falls, and higher death rates. Shari Schwanzl, vice president of operations and nursing for Helen DeVos Children's Hospital recommends that hospitals can keep compassion and prevent fatigue from setting in by making sure "employee-assistance programs are in place. The teams should come in twenty-four to forty-eight hours after a traumatic situation to help employees debrief and deal with the experience." Such employee-assistance programs are usually available, but the problem is the staff are not given the time to adequately utilize them. They should be mandatory after a crisis event with sufficient time for the employee to process what happened and adjust (Ermak, 2014).

Families

Health professionals should develop a greater appreciation for the factors and stresses that lead to burnout among families who provide the majority of ambulatory care to their loved ones. Comparing mothers of children with disabilities to mothers of children without disabilities, the former group experiences greater depression, anxiety, burnout, marital discord, and impaired physical health (Raina et al., 2004). Fathers are less susceptible to these effects. Mothers of children who could not reside at home have greater stress. Large families and long marriages experience poorer social functioning. Among geriatric caregivers, spouses experience greater depression, impaired cellular immunity, and a higher incidence of infectious illness. We need to understand the wide spectrum of disability and illness, the caregiver's emotional and physical health, coping strength, and socioeconomic support. This is basic to providing the best care for the patient and the caregiver.

Care for the Caregiver

Caring for the professional caregiver did not receive appropriate focus until the beginning of this century. With the shift to honesty and transparency in health care and the growing focus on how we could increase patient safety and quality, it became apparent that the very people providing the care at the bedside were being forgotten. The first step in providing this support requires health care leaders to create a just culture. In 1997, James Reason recognized that a just culture generates an environment of trust, urging and rewarding individuals for providing vital patient safety knowledge (GE Healthcare, 2011). Without trust, we cannot expect anyone to admit concerns or errors for fear of retaliation and loss of position. Once we establish the just culture, we must realize there is more than one person harmed in the medical community when something goes wrong.

Dr. Albert Wu from Johns Hopkins University first used the term "second victim" in 2000. According to Susan Scott, PhD, MSN, RN, "It's always been there, but all the pieces haven't been put together until recently. The effect of second victim trauma really gained recognition following the Institute of Medicine's 'To Err is Human' report in 1999."

The term has since been used to express the need to provide support to health care providers following an unintentional error or near-miss patient event. While the patient that the medical error affects is the first victim, the health care professionals involved are also victims. According to Dr. Wu, "'Second victim' health care providers have difficulty coping with their emotions after patients' adverse events with a prevalence from 10–43 percent."

Medicine is a privileged profession, but health professionals are not immune to emotional trauma. There are inherent hazards when caregivers fail to ask for help. According to Dr. David Mayer, the willingness to admit our humanness, and thus our ability to make an error, will help reframe the recovery and learning process when an error does tragically occur (2012).

Dr. Scott describes six stages of second victim recovery. The first three relate to realizing the impact of the event: 1) chaos and accident response, 2) intrusive reflections, and 3) restoring personal integrity. The latter three evolve over time: 4) enduring the inquisition, 5) obtaining emotional first aid, and 6) moving on. The challenge is that "moving on" tends to head in one of three directions—dropping out, surviving, or thriving. At the University of Missouri, Dr. Scott utilizes three tiers of second victim intervention. The first is local unit support by the manager/supervisor to connect with the involved staff, reaffirm confidence in their work, pulling in flex staff as needed, and regular supportive communication. The second tier of intervention comes from trained peer supporters and leaders in patient safety/risk management with one-on one counseling and team debriefing. The third tier of intervention is expedited referral to professional specialists. In the first three years of their care-for-caregiver program, 639 individuals received support: two-thirds in group briefings and one-third individually. Of the individual sessions, 15 percent were referred for tier-three guidance.

Current State of Affairs

The consequences of burnout and depression in physicians require further study. This topic is an important focus for several reasons. First, our students (medicine, nursing) and young physicians are exposed to these stressors and risk of experiencing the consequences. Second, they deserve leadership support to become comfortable in reporting their physical and emotional health. The majority of medical professionals in training avoid reporting such issues and seeking assistance for a variety of reasons, among them not wanting to appear weak. Third, the consequences are personal for the individual and for their patients. Of special concern is the potential impact on health care quality, safety, cost, and patient satisfaction. Fourth, the incidence and magnitude of these potential adverse consequences deserve continuing study and further clarification. Our students and residents are the generation that can lead this inquiry.

The medical profession has fostered an attitude and created a model as if it were a secret society performing initiation rites: "We will subject you to continuing stress and physical exhaustion, and only those who survive physically and emotionally are fit for the club." For too long, the medical profession has hidden and ignored the emotional and physical well-being of its members. Too often, this has occurred in ineffective and misguided peer review that seeks to shield struggling physicians from sanctions and/or licensure challenges. In addition, state medical boards have been found in violation of the American Disabilities Act (ADA) for seeking disclosure of the existence of mental health problems rather than compliance with effective treatment.

The American College of Graduate Medical Education (ACGME) periodically revises resident physician duty hours due to the impact of long hours on quality of care, patient safety, and resident well-being. This complex issue was the subject of an Institute of Medicine Report, "Resident Duty Hours: Enhancing Sleep, Supervision, and Safety," in 2009. Among the findings:

- Work Injuries
  - Percutaneous injuries were 2X greater at night than during the day.
  - Sharp injuries were 2X greater at the end of a long shift versus regular shift.
- Driving
  - The incidence of driving accidents was 2.3X greater after a long shift.
  - Comparing on-call residents to on-call attendings, residents experienced falling asleep at the wheel twice as often and had a 5.9X risk of near crashes.
  - Motor skills after long call are equivalent to having a blood alcohol content of 0.04–0.05g/100 ml (legally drunk, 0.08).

- Burnout
  - Overall incidence 41–76 percent with greater burnout in PGY1 (77 percent) versus PGY3 (42 percent).
  - There is a stronger association to burnout with work intensity as compared to sleep deprivation.
  - After the 2003 reduction in duty hours, there was a reduction in emotional exhaustion but no change in depersonalization or personal achievement using the MBI.
- Depression
  - Overall incidence of 7–56 percent with sleep deprivation being highly significant (7X greater).
- General health
  - Studies of sleep deprivation show a significant association with increased BMI thought to be mediated by changing levels of leptin and other hormones regulating appetite and less personal time for exercise.
- Professionalism
  - Sleep-deprived residents had 1.8X greater incidence of conflict with other members of the health care team and "didn't want to talk with families."

The ACGME is conducting a thorough revision of its recommendations on duty hours in 2016.

The studies cited above indicate that a disturbing percentage of medical students, residents, practicing physicians, and nurses continue to serve the public during periodic episodes of burnout and depression. However, the great majority of physicians and nurses seem to deal effectively with such stress. As seen in many groups, a small fraction causes the majority of problems.

Researchers in Australia report on the national experience with formal patient complaints to the regional health service commissions, civil courts, and the Medical Board of Australia, spanning 2000–2011 (Bismark et al., 2013). Of the 18,907 complaints, the majority were related to clinical care (61 percent) involving diagnosis, treatment, or medications. Nearly a quarter (23 percent) was related to poor communication involving attitude, information, and consent. Three percent of all physicians accounted for 49 percent of the complaints, and 1 percent of physicians accounted for 25 percent. There was a strong correlation between the number of complaints per doctor and the probability of future problems.

To a great degree, patients and families are unaware of burnout and whether or not anyone among their medical team was struggling. The avoidance of transparency can have two forms: that for which available knowledge is withheld and that for which further study is needed to clarify an issue. The latter frequently informs the former and makes sharing information valid and necessary. In defense of the opaque, sharing information that is preliminary, limited in scope or power, and needing verification can be detrimental. Misinformation can be worse than uncertainty.

However, transparency is a powerful force. Paul Levy, author of the blog post "Not Running a Hospital," said, "Transparency's major societal and strategic imperative is to provide creative tension within hospitals so that they hold themselves accountable. This accountability is what will drive doctors, nurses, and administrators to seek constant improvements in the quality and safety of patient care."

Lessons for Improvement

> Time waits for no man. Yesterday is history. Tomorrow is a
> mystery. Today is a gift. That is why they call it the present.
> —Alice Earle, 1902 (attributed)

It is popular today to speak of mindfulness—of being fully engaged in the present while having the emotional intelligence to understand the past and shape the future. As health care professionals, we are given an incredibly special gift by our patients and families. They trust us with their humanity, that we will not fail them in fulfilling our oath to provide the highest professional care, selflessly.

> I came to residency expecting to work as hard as I could
> possibly work, to get very little sleep, and to make very
> little money. These were prices I was fully aware of and
> willing to pay. What I did not expect was that I would

> work as hard as I could possibly work, and still do a bad
> job for my patients and my own education (Goitein, 2001).

Yet, there is reason for optimism as evidenced by two innovative GME programs. At the Brigham and Women's Faulkner Hospital, residents in internal medicine experience the Integrated Teaching Unit rotation. Patient census is capped at four patients per resident, and there are two teaching attendings on the unit, resulting in twice the exposure to bedside learning and four times the teaching. At Johns Hopkins Bayview Hospital, residents in internal medicine experience the Aliki Service where new admissions during the long-call are reduced from ten to five patients and during the short-call are reduced from four to two patients. Further, residents are expected to visit their discharged patient in their setting (home, SNF, hospice, etc.). Both the Faulkner and Bayview programs result in significant improvement in resident and attending satisfaction. According to the residents, "The Aliki Service provided us with a rare oasis in our training where we could practice the best medicine that we possibly could and rekindled the passion that brought us to our careers in the first place."

As we strive to meet our ethical obligation, it is meaningful that patients/families see our humanity, frailties, and fallibility. As a profession, we must remain mindful that all come to work with the best intentions; however, we function in a system and a culture that is imperfect. For these reasons, it is vital that we embrace just culture and engage care for the caregiver in its multiple facets.

Stories carry powerful incentive for empathy and accountability. We share the story of an anonymous health care professional who became the family caregiver:

> Although most marriages start with the words "in
> sickness and in health ... till death due us part," it is
> almost impossible to begin to imagine what it would
> truly be like to support and take care of someone you love
> and cherish through a terminal illness and death. Most
> families go through life with the normal ups and downs
> until they are suddenly confronted with an upheaval,

the likes that they could never have possibly imagined. How does it change the relationship when what was once a couple now becomes one spouse taking care of the other, when a parent is taking care of a dying child, and when a child is taking care of a dying parent? Availability for this type of caregiver is much more prevalent and well known. From cancer wellness centers to holistic health centers that not only treat the patient, but the families as well with social workers, massage therapy, support groups, etc. all available to help. While this can be beneficial to many, what unfortunately no program can do is take the place of the dying individual. Life as the family used to know it has been changed forever and will never be the same. The support is there for the family, but the loneliness is forever present. As Bono so beautifully says, "There is no end to grief … and there is no end to love."

Now imagine that you are the professional caregiver that was involved in the care of that patient. That you also bonded with that family. Worse yet, a medical error may have caused a premature death. All the support is given to the family, and you are left alone. If nothing else is important in this example, it is the need for care for *all* caregivers.

> Insanity: doing the same thing over and over
> again and expecting different results.
> —Albert Einstein (attributed)

We have an obligation not to accept the status quo. There remain remnants of the medical culture that attribute omniscience, infallibility, and unquestioned authority to physicians who are invincible against exhaustion and/or emotional instability. We cannot tolerate a culture that continues to subject its sons and daughters to physical and emotional trauma. We see that the consequences of failing to recognize everyone's humanity create burnout, depression, and suicide. Concurrently, we lose empathy, integrity, effective communication, and quality health care. We break our oath and lose our patients' trust.

Cultural change takes years—if not decades. The students and resident physicians who attend TTE are the future leaders of academic medicine, of private practices, of medical societies, of government organizations. Together, with a clearer vision of what is ethically necessary, with transparency, with humility, and with humanity, we can create the change that respects our patients, their families, and ourselves.

CHAPTER 11

Interprofessional Collaboration

Gwen Sherwood, Katherine Pischke-Winn, and David Mayer

Blog Posts Written by Health Care Learners

A Slow Nurse
Nicole Martin
June 18, 2014

Last night during a much-needed soak in the hot tub at Bear Creek Lodge, my solitude was pleasantly disrupted by a six-year-old fellow lodger. Amidst technical debates over water-gun engineering and musings on the existential dilemma of needing two hands to add up one's age, my new friend proudly announced, "And I just finished ski school … I can do the *pizza!*"

Ah yes, I thought, the ol' snowplow technique (forming a wedge with one's skis to slow and control one's descent down the mountain). I imagined the old familiar sensation with a smile. After the young man left for dinner with his family, I still could not shake the image of the snowplowing skier from my mind. I have not skied in years, but I realized that I had been living with the sensation of digging in my heels all throughout my previous year of nursing school and that this tension I had carried around all year was ... gone! And it was not just the muscle-relaxing magic of the hot tub. I realized that I had found a sympathetic community here at TTE and that I was feeling ... comfortable. Comfortable with myself as a *slow* student nurse!

Being perceived as *slow* in nursing school was a first for me, and though I felt it was a shortsighted characterization at the time, I also did not feel like I could argue with it. I *do* like to take my time. Each near miss or close call feels like a mogul to me, scattered around the field of my workday, and I carefully snowplow my way around each potential disaster. So many factors work against me in my pursuit of providing quality patient care. Surely, I should not tolerate being *rushed* to become the cause of a nursing error!

I do feel the pressure in nursing school. We students idolize the speed and skill of rapid-response nurses. When performing many of our procedures (such as catheter insertions of all kinds), swiftness is merciful. But I feel there are times to race and times to put on the brakes. I love it when a hospital unit I am working on is quiet, peaceful, and ... ahem ... *boring*. C'mon nurses, let's admit it ... *slow* is a four-letter-word to some of us!

Yesterday, John Nance paraphrased a nurse when he stated (referring to a cognitive strategy when dispensing medication from the Pyxis), "This is *poison* and going to kill my patient in five minutes if I don't double-check this!" The Three Checks, the Six Rights, and other procedures designed to slow nurses down are not glamorous parts of the job, but they save lives. What could possibly be more important?

What I am suggesting is a cultural change. Florence Nightingale wrote, "The very first requirement in a hospital is that it should do the sick no harm." I was involved in three good catches last semester, all of them clearly related to the need for others to slow down. High acuity and large patient workloads are indeed very real issues that constrain our time, but my observations lead me to believe there are other factors at play such as a cultural norm of speed, often at the expense of safety (for the safety of the patient *and* often the nurse!).

Not *slow—safe*!

Gang Warfare
Evan Skinner
August 1, 2014

We often only see our input to patient care as an item of control. John Nance's book for the conference, *Why Hospitals Should Fly,* discusses this very point. It is often surprising and sometimes challenging to physicians when nursing is able to keep pace with a discussion of complex pathophysiology or pharmacologic interventions for a given presentation. The knowledge of that which constitutes a diagnosis as described per a nurse should not be viewed as an interloper moving against the castle. Would it not be beneficial to have someone monitoring a patient who has a deeper understanding of patient care than is expected of them?

As well, do nurses not demonstrate intimidating and humiliating behaviors with others in their own peer group? Bullying is a very real phenomenon in nursing. An example of nurse aggressiveness is the following: Other staff entering the rooms of patients assigned to certain nurses are treated as burglars entering a private residence rather than someone providing additional vigilance for a patient. I've witnessed nurses telling physicians, especially residents, to enter their patient's room only with the nurse's knowledge and permission. We nurses are notorious for eating our young, certainly within the ED or ICU. Often, new nurses entering these arenas are humiliated or "pimped" as a consequence of their preceptors' own experiences being made to feel insignificant earlier in their careers. The condition of transference of violence is one of the greatest single impediments to the advancement of nursing.

Health care in some respects has deteriorated into turf battles in which all the professionals are scrambling to grasp the largest amount of control or influence. Let's move beyond psychological shackles and rule our practice with objectivity. Let's understand that we all share the same goal of 100 percent safe and effective patient care and each professional having an impact to provide for the common goal. Let's not separate ourselves with razor wire across a demilitarized zone. Reach across the aisles, shake hands with others, and begin the cease-fire.

Communication and the Optimism Bias
Nicole Werner
July 31, 2014

Communication, teamwork, and *team coordination* are buzzwords for patient safety and improved health care system performance. But does everyone truly know the meaning of these terms? We were presented a case study today in which an experienced and revered surgeon encouraged his surgical team to break protocol by closing a patient when the sponge count was off by one sponge. The surgeon cited, with amiability, the length of time under anesthesia as the urgency for ending the procedure before the missing sponge was accounted for—and rightly so since longer times under anesthesia are associated with decreased patient outcomes. However, so are surgical materials left in patients. The concerned team deferred to the surgeon's congenial and persistent request for the sutures to close the patient.

At first glance, this team could be considered to have some of the above qualities: they communicated about the missing sponge, they coordinated a course of action, and they did so in a seemingly patient and kind manner. However, my view is that the surgeon manipulated the team using his knowledge of their reverence and his kindness as a tool to encourage the response he desired—not to intentionally harm the patient. Certainly, he was acting in what he thought was the best interest of the patient as the decision-maker for that patient's care. Sometimes, being too close to your team, in the absence of a structured communication method, can be dangerous.

The issue is, even if there had not been a sponge left in the patient, this is still a critical breakdown in communication and teamwork. The team leader encouraged the team to break hospital safety protocol, and the team followed along. No one persisted to communicate the problem. The team did not come together to solve the issue; instead, they relied on one person's decision. Health care providers tend to have an optimism bias when it comes to serious safety events; they typically do not believe that bad things are going to happen to them. And surgeons especially require a certain level of confidence to perform their jobs. The problem is that bad things, like accidentally leaving a surgical sponge in a patient, do happen. All the time. And the protocols are there to protect against this. Encouraging the team to break protocol endangers the patient as well as the whole team. Although this event was multifaceted and would require a thorough evaluation (and full disclosure of harm to the patient and/or the patient's family), two system-based improvements could help:

- There is a need for communication standardization and training so that nurses feel that they can persist against the wishes of the physician—and have the training to do so—and that the physicians have the training to accept feedback from the nurses with whom they work.
- As discussed throughout the past couple days, there is an overall need for error reporting transparency within hospitals (at least) and ideally across hospitals so that health care providers can gain an awareness of the frequency and severity of serious safety events and reduce the optimism bias.

We traveled to Arlington Cemetery today, a place I have been more times than I can count. It is always a requested tour of visitors to my home in Washington, DC. Today I saw this space in a new light. Four-hundred thousand was the number. Four-hundred thousand graves of soldiers who have given their lives in the name of country and freedom. We could see many of them from our perch on top of the hill. Four hundred thousand gravestones spanned the space below where we stood; today, that represented both the soldiers who gave their lives and the approximately four-hundred thousand people who have died due to preventable medical error. Those errors are like the one described above, which many health care providers do not think can happen to them.

Initial Reflections

These three student reflections share dominant recurring themes of any safety culture: leadership, communication, and teamwork. Each instance reveals gaps in our systems of care that are barriers to creating effective safety cultures. When nurses feel pressure from task overload because of complex patient acuity or workarounds to overcome too few resources, safety suffers. When management values efficiency over value and effectiveness, safety suffers. These stories call for system repair; in fact, they demand system redesign to alleviate workplace barriers while creating an environment where patient safety is *always* the first priority. The reflections highlight the role of leadership in role-modeling transformation and interpersonal interactions that promote teamwork and collaboration rather than turf management, as the second writer comments. Taken together, these three reflections challenge our traditional view of health care organizations in hopes that we begin bringing a resilience science system's approach to all we do. But how can we reshape leadership, communication, and teamwork to see the interplay they create in building a safety culture?

Safety culture demands effective communication that shares critical information and uses the best evidence to make informed decisions regardless of who sends the message. Communication is the very heart and soul of culture. Communication and culture are intertwined; culture is expressed through the social interactions of group members. Clear, assertive, honest, and trustworthy communication between caregivers within a nonpunitive environment ensures transparency and keeps everyone on the same page to avoid harm.

The success of interprofessional teams will be highly influenced by their application of high-reliability organization (HRO) principles and by emotional intelligence. They are the threads that bind leadership, communication, and teamwork together to create high-quality, safe patient care and can help bring caregiver joy back into the workplace. The five key principles of HRO are: preoccupation with failure, reluctance to simplify, sensitivity to operations, commitment to resilience, and deference to expertise. Emotional intelligence (EQ) is a critical skill for leaders and teams to promote interprofessional practice. It is comprised of four domains: self-awareness, self-management, social awareness, and relationship management. We will visit facets of HRO and EQ later in this chapter.

Historical Background

Historically, health care providers operated in steep hierarchies in which physicians made all the decisions and led care delivery. As the dynamics of health care changed over the past decades with more complex patients, exploding health care information, and developing technology, no one discipline can be responsible for an individual patient. Having the best outcomes requires shared information and shared decision-making among providers and the patient and family as well.

In the past, the example of the retained sponge or any retained object during an operative procedure would have been the sole responsibility of the physician. Now, with careful attention to detail, sponge counts, and other standardized procedures enforce checks throughout the process so that situations like awareness of lost sponges should be communicated in real time instead of the later stage of the operation. Still, there remain issues with communication. Nurses may not have been taught assertive communication to speak up and share critical information, especially if it relates to an error by a team member—or, in this case, a preeminent surgeon. Physicians may have been trained to be the "captain of the ship" and not understand that responsible leaders welcome input from all team members. Communication is still too often driven by traditional hierarchy in which the halo effect—when no one will question the expert—prevents speaking up.

Since the Institute of Medicine's "To Err is Human" report in 1999, we have known that communication accounts for roughly 70 percent of health care harm. At the heart of transforming the health profession's education is learning to recognize that essential information must be relayed to a team member *and* developing the communication skills to effectively convey that information. In 2003, the Institute of Medicine report called for radical transformation of how we educate health professions and identified six competencies students achieve: patient-centered care, teamwork and collaboration, evidence-based practices, quality improvement, safety, and informatics. Academic institutions that teach health disciplines still struggle with uneven implementation and curricular integration. The Interprofessional Education Collaborative has further developed four domains that are essential for working together across health care, which are adopted by all major health care professions: roles and responsibilities, ethics and just treatment, teamwork and collaboration, and effective communication (IPEC, 2011). The Joint Commission has issued regulations to ensure hospitals develop policies and procedures to address bullying and incivility.

Still, we have examples of provider-versus-provider behavior that jeopardizes the well-being of patients and undermines future teamwork. Nurses treat each other without respect, physicians berate nurses, and management mismanages workers. Recognizing the role of culture in redesigning systems is a first step toward changing provider knowledge, skill, and attitudes about the role of safety, teamwork, and collaboration, patient-centered care, quality improvement, evidence-based practice, and informatics. While we can agree on the knowledge and skill development, changing attitudes is a complex process. In the next section, we will explore how reflecting on our experiences can help open our minds to reexamine how we interact, communicate, and work together to keep patients safe.

Current State

From Reflection to Transformation

Narrative, as told through stories and exemplars, has been a powerful and effective tool for creating change. Stories from practice weave organization theory and experience through a reflective process that shares the transformative process (Powell & Stone, 2015). Story helps us see how change leading to transformation came to be and its impact on culture. Story can help shift organizational culture by putting a lens on attitudes and behaviors that drive actions and interactions. This is particularly true in examining the reflective stories that lead into this chapter on nurses and physicians and how they approach their work together.

Reflective practice is the thread in transforming organizations; the stories in this chapter are the reflections from nursing and health science students participating in TTE. Reflective practice is thinking consciously about one's actions and responses. Systematic thinking pulls in knowledge and previous experience in light of one's values and beliefs to begin a reframing process to develop future responses or actions. Reflective practice is closely related to emotional intelligence: developing self-awareness to be conscious of the impact one has on another and how that drives interactions and behaviors.

Organizational culture is the collective values, beliefs, and norms held by group members (Triolo, 2012) and thus can enhance and reveal dysfunctional patterns. Learning organizations focus on improvement and operate with congruence of values and actions among the leaders and workers at all levels and microsystems. *Culture* itself is the interplay of structure, the people themselves, leadership, reward systems, and processes and interactions through which the organization gets its work done (Triolo, 2012). Those in the organization display facets of motivation, trust, and communication in how they interact and accomplish the work of the organization (Schein, 2010). Culture is learned and reinforced by what members see, the behaviors they observe as successful, and what is tolerated by those in the group. If good behavior is unrewarded and bad behavior is tolerated and not punished, there is distrust and dysfunction within the organization. Therefore, culture is shaped by how work is accomplished and how problems are resolved. Actions are reinforced from witnessing behavior in critical events modeled by the leader. Members assume that these behaviors work; therefore, they become part of the assumptions underlying the organization's culture.

We witness organizational culture in the way that staff deliver patient care and establish relationships up and down the organization; relationships reveal informal values, beliefs, and norms held by those in the organization. Group members easily become entrenched in established behaviors and communication patterns that are difficult to change. Thus, organizations have a difficult challenge to adopt new behaviors and attitudes to become safe, reliable organizations. A distinguishing characteristic of healthy organizations is the capacity to continuously improve and adapt to change and innovation (Schein, 2010).

Communication is an essential organizational process and shapes culture. Culture and communication are intertwined such that culture is lived through the interactions of its members (Bellot, 2011). Group members receive, interpret, and evaluate input through continuous communication from and by the organization using multiple approaches. Organizations are changed and sustained through how the culture manages change processes, shares stories, resolves problems and conflict, and approaches teaching and learning.

Shared mental models are the overlapping concepts held by group members and include the thought processes about how something works in the real world. They help get everyone on the same page, which lessens ambiguity about what is acceptable in the organization through shared representations of events and concepts. Shared mental models provide a framework to assess and explain what happens in the organization's environment (Van den Bossche, Gijselaers, Segers, Woltjer, & Kirschner, 2011). Group members develop shared mental models through explanation of thought processes and working relationships. Shared mental models enhance team communication and effectiveness in completing complex tasks; therefore, a closely held, shared mental model enables the system to learn and adapt to rapidly changing environments and achieve best outcomes.

Understanding how culture is formed and reinforced throughout the organization is essential for creating a culture of safety. Assumptions are deeply held and difficult to change, yet uncovering assumptions and advocating for change with its inherent uncomfortable "feeling" is the starting point to change culture. The beliefs, values, and expectations of the organization's members affect organizational change. Learning from experience involves action and reflection. Through reflection, concrete experiences are assimilated and transformed into abstract concepts and become the basis for actions that are tested in new experiences (Kolb et al., 2000). Learners experience, reflect, apply, and evaluate their experiences to try to rethink how they will respond in future similar situations.

Culture is revealed by the attitudes and behaviors of individuals. In promoting communication and teamwork, consider the HRO principle of deference to expertise. Who is the expert? In multidisciplinary team rounds, the attending physician is the medical expert and team leader. Yet, he/she is not the nursing expert, or social work expert, or pharmacy expert. Indeed, who is the patient expert—if not the patient and family? Therefore, it is essential that the physician respects and includes all team members for their expertise during collaborative rounds, especially the *patient.* Expertise is not defined by the string of letters following a name—MD, MBA, FACC, FACP. Instead, it asks who knows the most about the specific topic of conversation.

Furthermore, the leader's ability to be inclusive and respectful requires emotional intelligence (EQ). Team success is more dependent on EQ than IQ. Authentic leaders understand their strengths, weaknesses, distractions, hot-button triggers, and ethical obligations while seeing the current task in the context of a larger vision. They are self-effacing, transparent in sharing information, complimentary to each team member, and compassionate toward the patient. Moreover, they perceive how these characteristics operate in others. In so doing, they influence, mentor, inspire, resolve conflict, and earn the trust of the team, patient, family, and organization.

If we are to change organizational and safety culture, we have to examine our own attitudes and behaviors and the assumptions, values, and beliefs that guide them. Reflective practice can challenge our traditional responses so that we see benefits of new behaviors. We see that if we had been assertive enough to speak up when we know a sponge is missing, speak up when patient care is being compromised, or speak up so that we do not tolerate incivility, we can improve outcomes. Safety belongs to all of us, but it begins with the individual—and the competencies each person must develop to ensure safe care for every patient every time.

Lessons Learned

In exploring the lessons from the three reflective stories from nursing and health science students, we have reviewed the role of organizational culture in safety and how leadership, communication, and teamwork are essential domains to develop safety culture. Reflective practice is a structured examination of experience to reframe what we know, what we need to do, and how our attitudes and assumptions shape how we act. Culture is driven by communication; the way members communicate is at the heart of how organizations achieve their goals. Members need to develop skills for clear, concise, and trustworthy communication to change interactions and develop reliability across microsystems.

Successful interprofessional teams require leadership, but leadership is not determined by hierarchy. Implementation of high-reliability principles and development of emotional intelligence create a space where *any* member of the team can be a leader. Organizational success is not determined by a chart of titles but by a network of respectful relationships that create collaborative teams, including the patient and family.

Transparent communication is a major focus of TTE. Low-fidelity simulation gaming strategies have been an engaging learning approach that helps achieve behavior changes. To conclude this chapter, we describe the simplicity of using dominoes to help develop communication and teamwork skills and increase self-awareness of the impact our interactions and communication have on outcomes.

Providers want to work in organizations that focus on safety and quality, develop reliable systems of care, and foster positive, team-oriented work environments in which members demonstrate trust, respect, and value for each other. The view of white grave markers in Arlington National Cemetery seems endless ... the impact of that same death toll comparison to the epidemic of preventable medical deaths each year is staggering. It is time for all of us to take action to change the culture in our systems of care and demand leadership, communication, and teamwork behaviors that will keep patients safe—every one, every time.

CHAPTER 12

Positive Examples
of Culture
Lisa Freeman

Blog Posts Written by Health Care Learners

A Small Step toward a Just Culture
Mihai Dumitrescu
August 4, 2014

When David was talking about just culture on our last day at Turf Valley—and empowering every member of the team to speak up without fear—I remembered a story.

As pediatric residents, we rotate in the newborn nursery. The morning routine is to ask the nurses to collect the babies at five forty-five to get the blood for the routine bilirubin checks and newborn screens. We start examining the babies before six o'clock. Many times, I help nurses collect the babies and explain to the new mothers what is happening with their newborn infants.

I have a system I follow daily after aligning all the infants in the room with big windows on the third floor. I start examining all the babies one by one, and I go through each system, making sure I finish with the red-light reflex on each one of them. When the attending arrives, we discuss the patients. I ask all my questions and then complete my daily notes and discharges for the day. By then, it is seven thirty, and the infants—with new blankets, new diapers, and often funny hairstyles—are ready to go back to their mothers' rooms. One person—attending, resident, nurse, or other ancillary staff member—should be present at all times in the room where we collect, examine, and change the babies. I enjoy talking to the new mothers, explaining what to expect from their newborns, and answering a multitude of questions regarding the feedings, signs of hunger, activities, sleeping, when to take the baby outside, when it's safe to go to the beach, vaccinate, following up with the pediatrician after discharge, or whatever questions new, anxious parents might have. I often find myself taking the infants back to their rooms.

I'm always smiling, and I really like to get to know the members of the hospital team. When I go to discuss plans with my patients, I know the names of every person who walks in the room and interact with them—from the charge nurse to the lady who brings the meals.

One day, a nursing assistant was in one of the rooms when I walked in. Maria welcomed me with a big smile, telling the mother how lucky she was to have me as her doctor. I was very happy to hear that, and at some point as I was focusing on the mom to see if she had additional questions for me, Maria left the room, though I don't even remember when. Once I finished with the mom, said goodbye, and wished the best to the family, I left the room and found Maria outside the door.

With her hand on my left shoulder, she said, "Doc, I would not say this to any of your colleagues, but I really wanted to say something to you … may I please give you some advice?"

For a moment, I didn't realize what she was talking about, and I felt my ears getting hot and red. I tried to smile, but I couldn't. I was nervous and curious.

She continued, "I saw you bringing the baby to the mom in room 385, and, Doctor, just some advice … don't get upset!"

At that point, I really felt uncomfortable. I knew I had done something wrong. "Doc, you didn't check the bracelet on that baby, and you didn't ask the mother for her ID band. I just wanted to tell you this because, some time ago, one of our nurses got fired after she took a baby to the wrong mother, who breast-fed this infant."

I realized in that moment how important that step was, and even though it sounds normal to check, that was not part of my routine. I gave her a big hug and thanked her. She was the first one to tell me that—even though she was not the first one to see me taking babies to the mother's room. In that moment, I realized how close I had been to making a mistake the day before. The room number was wrong on a baby's bassinet, and while walking in the room, I saw the mom with another baby (her own) at the breast. I was confused until nurse told me they had switched my infant's mother to a different room and directed me to the right place. I realize now that, even when I was so close to making a mistake, I still didn't check the bracelet. I realized what an important lesson Maria had given me …

I asked my fellow residents that same day if they checked the bracelets on each baby and their moms, and all of them said, "We should." However, nobody was doing it because nobody told us at the beginning of the rotation about that step. It was Maria who changed our routine. I wonder how many mistakes we are avoiding every day by checking ID bracelets, and it is all because of Maria. She probably doesn't even know the term *just culture*, but she knows to speak up when something is not right or when procedures are not followed. She spoke up because she knew I would appreciate it, but to make it a culture change, we have to be open.

This is a big step toward just culture, and I wanted to share this story with you because I really hope there are many Marias out there. They are the key team members who can be used as role models in our efforts to become open-minded and honest on our way toward zero preventable harm.

A Culture Change in Medicine: From Do No Harm to Do Less Harm
Cassidy Dahn
June 14, 2015

We all cite the Hippocratic oath when we enter medical school. I remember like it was yesterday, standing together, holding up our hands in our short white coats, and swearing to follow its covenants throughout our careers.

To my surprise, as I referenced the oath today, the phrase I remember most—"First do no harm"—is actually not part of the oath. Somehow, that is what I remember holding my hand up and promising to uphold.

This is the culture of medicine that is ingrained in me, changing my memory and perception of what I swore to that day. The culture that is accepted in society, among doctors, nurses, and you and me, is that doctors are infallible. It is not fathomable at times and is certainly not understandable or acceptable for physicians to make mistakes. But physicians are humans, and humans make errors. There are unreasonable expectations that breed more and more unsafe practices, lack of transparency, and lack of learning from our mistakes.

We ended with a video today that closed with the quote: "Do less harm." That quote was one of the most thought-provoking for me. It spoke to the many lessons I learned and jotted down on my notepad during the day about patient safety:

- Doctors are human, and doctors (including you and I) make mistakes.
- Avoid cognitive bias/anchoring/whatever you want to call it. It's deadly. (I don't understand how we prevent this easy-to-fall-into trap.)
- Learn how to balance mindfulness and wastefulness.
- Learn how to be reassuring but always continue to reassess.
- Say I'm sorry, say it early, but don't stop there.
- The patient and family, no matter what, is *always* right.
- Errors need to be recognized, reported, and appreciated. It is unrealistic to promise to do no harm, but it is commendable to promise to minimize the harm in our unsafe system that exists and do *less* harm.

This is not to say we shouldn't aim for a zero-error rate—we all absolutely should—but we need to be more accepting of human (including caregivers') fallibility.

Culture and Language
Colleen Parrish
June 9, 2015

129

Language is a fundamental piece of any culture, and within the health care culture, it is no different. In order to begin change within the health care culture, we first need to adopt a change in our language: the language we use interprofessionally, the language we use to describe patients, and the language we use to interact with patients and families. In changing, or perhaps evolving, our language we will improve our communication and our understanding of the human nature we all possess.

I reflect back on a big movement within the special education teaching realm to change the language used to describe a patient with disabilities. Of course, we have transitioned from using the term *mental retardation* to *intellectually disabled*, but more importantly, there was a change in the *sequence*. It is no longer "that special needs kid" but rather "that kid who has special needs." It was a drive to put the child or person first in the language we used in order to reinforce in our minds that, in the end, these are people too, and they really aren't that different from the rest of us. Changing the language we use changes the way we think about people and situations. I reflect on a simple transition I have personally tried to make from saying, "I *have* to do something" to "I *get* to do something," and it has really changed the way I approach situations and puts into perspective the simple things I took for granted.

I think it's great to see the start of this in changing "calling for help" to "calling for guidance." It's just one word, but it creates a completely different mind-set and starts to form or *evolve* a culture.

Culture Change? Which One?
Garrett Coyan
June 21, 2013

The last week I spent at Telluride was very eye-opening for me. I was glad to be surrounded by so many other health care professionals who had the same desire to provide the safe and high-quality patient care experiences as I do. Reinvigorated with ideas for improving communication and decreasing risk to my patients, I couldn't wait to get back to my institution and start implementing change. However, as I returned to the hospital today, I was quickly reminded of the main reason why this goal will be so difficult. Not only does cultural change need to occur in the hospital, but I would argue that even more importantly, cultural change needs to occur in the education of health professions students. This was made evidently clear by a conversation I had with one of my recently graduated colleagues who is starting his internship in a week.

At our school, we take a one-month class in our fourth year about public health, health policy, and health care practice. A few days of this class are slated to discuss patient safety and quality improvement. When I asked my friend about the content of this lesson, he told me frankly he couldn't remember because he and most of his classmates thought the information was either common sense or not really applicable. He related that most of this lesson was simply going over "bundles" of different types—and maybe fifteen minutes was spent on communication between providers. The material was never really tested, and apparently, it didn't stick with this particular colleague of mine. This, of course, was rather startling to me after the experience I had just had in Telluride!

In order to change the safety culture of the hospitals and health systems we work in, we need to create providers who are trained, knowledgeable, and willing to implement the changes needed to provide quality, compassionate, and safe care to our patients. It is for this reason that I will be speaking with my dean, the director of quality improvement, all of my mentors, and as many of my classmates who will listen about making this information a mandatory and testable portion of our curriculum. It is only then that we can hope for young medical students, residents, and physicians to be competent and comfortable enough to speak up when medical errors are made and confront them head-on and honestly with our number one partner in health care: our patient.

I suppose the hardest part of the Telluride experience was not being involved in the intense and productive discussions that took place but coming down from the mountain (literally and figuratively). It's time to get to work. Culture change will not occur until we start demanding it!

Initial Reflections

Culture: The integrated pattern of human knowledge, belief, and behavior that depends upon the capacity for

learning and transmitting knowledge to succeeding generations; the set of shared attitudes, values, goals, and practices that characterizes an institution or organization *<a corporate culture focused on the bottom line>;* the set of values, conventions, or social practices associated with a particular field, activity, or societal characteristic
—Merriam-Webster

What is the culture of health care and medicine? The answer is not obvious, simple, or straightforward. Based upon what we know, the culture of health care is changing. In the blog posts that are included in this chapter, we hear four different themes that allow us a glimpse of the direction that culture change is moving in: Patient safety benefits when we accept that doctors are fallible and that they are only human. When health care professionals feel safe to speak up without fear of retribution, medical errors and patient harm are reduced. The language we use changes how we approach situations and people. And to change the culture within health care, we must start teaching new ways of thinking early—in medical schools and other educational programs.

Many of our medical systems are still quite traditional, and the culture under which they operate is based on a strict hierarchy with the physician at the top and the patient at the bottom of the ladder. In these systems, many of the doctors feel that they must always be right, and as a result, communication flows in one direction. This lack of a team approach sets up the system for repeated errors and harm and an impaired ability to learn from the mistakes so that they won't be repeated.

In the blog post, "A Culture Change in Medicine: Do No Harm to Do Less Harm," the author speaks about her early impression that the culture of medicine included "First do no harm," which she recalled as the central part of her oath. The absoluteness of this edict, which is not actually part of the Hippocratic oath, is loaded with implied pressure to never make a mistake. But people make mistakes, and people in health care will also. To improve the culture of patient safety, it is necessary to understand that errors can happen and then learn and teach strategies to reduce their likelihood. This approach recognizes that health care must also include the patient and their family in their care, and if an error occurs, the system must promptly acknowledge it, apologize for it, appropriately compensate the patient, study it, and learn from it. This blog post sums it up at the end: "It is unrealistic to promise to do no harm but commendable to promise to minimize the harm in our unsafe system that exists and do less harm."

In order for there to be an effective culture of patient safety, there needs to be a culture that allows and encourages everyone to speak up without fear of negative consequences when they see something they believe could lead to an error or patient harm. The adage, "If you see something, say something" applies here. In many instances, errors or harm occurred began with something that someone noticed but didn't act upon. Perhaps they did not realize the potential or perhaps they were afraid to say something.

Just culture supports and promotes an environment where, in the interest of patient safety, every member of the health care team feels safe in speaking up without fear. In the blog post "Small Step Toward a Just Culture," we appreciate how when one employee felt safe in speaking up, the result was that all of the residents changed their practice, thus avoiding an avoidable event. Just culture allowed for a learning moment. Without a just culture, health care is defensive and secretive. Errors are not openly discussed, and mistakes are bound to be repeated. With the sheer number of medical errors resulting in making it the third-leading cause of death, it is critical that our hospitals create a supportive environment where everyone feels a commitment and a responsibility to speak up when they become aware of something that already has or that may put patients at risk.

"Culture and Language" speaks to another cultural change that began in the world of special education, but it speaks to health care as well. Patients who are involved in the health care system are people first. They may happen to have a medical condition that they are being treated for, but that is only a part of who they are. They are not "the heart attack" in room 205; they are "Joe, who had a heart attack." Their being should not be defined by the condition that they have but rather by who they are. The language that is used sets the tone for the entire health care experience and should be carefully chosen and purposefully used.

In large and small systems, our blog post, "Culture Change? Which

One?" asks where this cultural change begins. It can begin wherever there is a champion who will lead the way. Within a hospital system, it can be through a policy that supports a culture of patient safety in all aspects of the health care they provide or it can be through small changes that occur in a single unit.

"Educate the young and, when necessary, regulate the old." Following this mantra, another important place to begin is in the classroom. Students are tomorrow's leaders. If they are taught and understand concepts of just culture and person-centeredness while they are still in school, they will be empowered and equipped to be the champions who can lead their health care systems into culture change that will improve patient safety and the quality of health care.

Culture Change: What Was It Before?

Years ago, when patients were harmed in the course of receiving health care, the majority of them were told very little and had only one place to turn to get answers and find out what really happened: the court system. According to *Modern Healthcare*, "Surveys have shown that patients and their family members are far more likely to pursue legal action when they feel a lack of accountability, disrespect, or poor communication from their providers" (*Modern Healthcare*, 2015).

The policies that many hospitals established left patients and their families feeling empty, confused, and frustrated. Systems were not transparent; policies in place advised employees not to reach out to patients, not to apologize, and not to engage in detailed conversations. Even more importantly, the wall of secrecy did not allow providers to learn from the errors that occurred, and little was done to prevent them from being repeated. Health care employees were not encouraged to talk about patient safety issues that concerned them. In fact, there are many anecdotes documenting situations where a newer or "lower-level" employee observed a physician making an error and did not feel comfortable speaking up and saying something. Situations that had the potential to lead to patient harm often did.

In November 1999, the Institute of Medicine's "To Err is Human: Building a Safer Health System" found that there were between forty-four thousand and ninety-eight thousand deaths each year as a result of preventable medical errors. At the time, this finding surprised many. This increased awareness of American medical errors brought reflection about ways to reduce patient harm. The realization started to hold that learning could lead to a reduction in errors if it were given more openness and transparency. But older patterns and habits are hard to break, and years went by with only small system changes and a lot of responsive legislative actions on tort reform.

In 2010, the Affordable Care Plan was approved and signed in to law. It includes various financial incentives to reduce medical errors, which has led to changes in the culture of patient safety. By way of example, hospitals are no longer reimbursed by CMS for all the separate services that are associated with a surgical procedure. As a result, when harm occurs during a surgery that requires additional surgeries and procedures, the hospital is reimbursed the same as if no complication occurred. As a result of this and other changes and incentives to reduce harm, systems throughout the country are researching and implementing ways to improve patient safety, improve the patient experience, and reduce health care costs. This established the framework that the Institute for Healthcare Improvement (IHI) calls the *Triple Aim*, which describes an approach to optimizing health care system performance.

In 2013, John James, a retired NASA scientist whose son died following avoidable medical error, researched medical harm leading to death in the United States and established that it was much more prevalent than had even been realized in the IOM report of 1999. His results were published in the *Journal of Patient Safety* (*Journal of Patient Safety*, 2013). He established that between 210,000 and 440,000 people die each year as a result of medical harm, a figure that has now been widely accepted.

Following these developments, we started to see a change in the culture of patient safety with new and different approaches to patient safety and quality improvement. The change is complex and has been slow in taking hold. We see some systems embracing a new way of approaching patient safety and health care delivery, others doing business as usual, and the majority making efforts to change and falling somewhere in between.

Several years ago, in places like Michigan, Oregon, at the VA, and a few other locations, discussions began to focus on various communication and early resolution programs. Not all of them offer the same degree of consideration and assurance of an impartial and fair resolution to patients who have been harmed, but the conversation has begun to take place—and so has the culture change.

Today, more individual hospitals and additional systems have continued to embrace some of these as well as other programs. An important component of many of these programs is the concept of just culture, which has been discussed in this chapter. Equally important is the rise in awareness of the patient perspective and the inclusion of patients in all aspects of health care delivery as equal members of the health care team—from the exam rooms to the boardrooms. Through legislation, understanding, and education, the culture of health care is continuing to evolve, and it is up to the individuals involved to embrace and perpetuate the changes at the local level.

Current State of Patient Safety Culture

The culture of patient safety is evolving as this is being written, and overall, it seems to be moving in a positive, though not terribly consistent, direction. I would like to say that the current culture trend of patient safety and health care in general is one that is recognizing a more progressive approach to health care—one that does focuses on treating sickness and on promoting health and safe, high-quality care. It follows a patient/person/family-centered philosophy and looks at the language of health care through a new perspective. All these strategies involve cultural change that focuses on broad communication and teamwork strategies.

According to Grissinger, "During the past decade, those most involved in the patient safety movement have come to realize that preventing catastrophic events—or any avoidable harm to patients, for that matter—requires more than simply changing systems and implementing best practices" (Grissinger, 2010). However, while this is being realized more broadly and accepted by many, the necessary changes for this change in culture are not easily implemented, and often those attempting to do so find it very challenging.

According to Bryan Sexton, associate professor and director of the

Patient Safety Center for the Duke University Health System, in a 2013 interview, the improvement in the culture of patient safety over the prior seven years has been spotty and inconsistent. He pointed out that while teamwork and person-centered care are recognized for their importance in improving patient safety, innovation fatigue leading to burnout was occurring in significant numbers. He believes that when the burnout rate is under control, then the staff become resilient and are prepared to work on issues around teamwork.

As improved patient safety and patient-centered care are both predecessors as well as goals of culture change, we have to do much more deliberate work to support health care workers in reducing burnout and increasing resiliency. With that in place, process improvement and changes can take place, and the cultural changes needed for sustained improvement in patient safety can occur (Wachter, 2013).

Recognizing that the challenges are many, we see that the positive culture of an organization is correlated and predictive of its success in keeping patients safe (Smetzer & Navarra, 2007). While it is recognized that it may initially be difficult to change the way that care is delivered and it comes with its risks that the changes may not work as expected or that there may be unintended consequences, there are locations around the country and internationally that have embraced many of these forward-thinking principles and are seeing positive outcomes. The results of these changes can be observed through the use of various tools including the AHRQ Hospital Survey on Patient Safety Culture tool and resulting annual report. From this particular survey, hospitals are able to see how their patient safety culture compares to established benchmarks and to other hospitals. From the 2014 Agency for Healthcare Research and Quality report, we see that perceived areas of strength were:

- teamwork within units (81 percent positive)
- supervisor/manager expectations and actions promoting patient safety (76 percent positive)
- organizational learning—continuous improvement (73 percent positive)

The areas perceived as needing improvement were:

- nonpunitive response to error (44 percent positive)
- handoffs and transitions (47 percent positive)
- staffing (55 percent positive)

Aside from these surveys, one can also see how hospitals are doing in terms of their patient safety achievements by looking at some of the hospital (and other health care) rating sites. While there are various aspects of care being measured by the various tools, several focus on patient safety, including Leapfrog Hospital Safety Score, *Consumer Reports*, HealthGrade's Distinguished Hospitals for Clinical Excellence, and Medicare Hospital Compare. What is seen confirms what was stated by Bryan Sexton: there is more inconsistency than not amongst hospitals (Leapfrog).

Despite the powerful stories of improving and high-performing hospitals, improvement across the board remains elusive. According to Leapfrog, the fall 2015 update shows a number of positive trends for certain hospital-acquired conditions and safety measures, but hospitals are performing worse on critical measures like foreign objects left in after surgery. Overall, performance on safe practices and process measures varied greatly. It should be noted that among the 2,530 hospitals that were scored by Leapfrog, 773 earned an "A" and only 34 a "D," reflecting a leaning toward better patient safety. Quite importantly as well, 133 of the hospitals earning an "A" have done so since the Safety Score began in 2012, and a number of others showed improvement (*Hospital Safety Score*, 2015). There is still work to be done in the quest for greater patient safety.

Lessons Learned

Imagine a world where safe, high-quality health care is provided to all patients all the time. Picture a world where all of health care puts the patient at the center of the thought processes and where all of health care, at every level of care and policy, includes patients as partners. This is the culture of patient safety in health care for which we are striving.

As we have talked about in this chapter, it is a culture that is aware of, and sensitive to, the person who is a patient and appreciates that there is more to a patient than their current medical condition or illness. It is a culture that recognizes that each person is an individual and that their preferences must be heard and respected. This culture recognizes that patient safety and person-centered care are intrinsically interwoven. It is also a culture that recognizes and accepts that doctors are fallible, as are all people. It provides for an environment where all health care professionals within a health care system, from the CEO to the parking attendant, are safe to speak up when they see things that can be improved or when they have seen errors that have occurred. It is a culture built on mutual respect. Finally, systems that are successfully making changes in their culture of patient safety are also making use of high-reliability strategies and are learning from other hazardous industries, such as the nuclear and airline industries, which have experienced relatively few safety events. While we are changing the way things are being done in our health care systems, in order to fully achieve this culture change, we must also begin educating our future health care leaders—medical students, nursing students, and students of all the health sciences in these contemporary ways of thinking and approaching health care throughout their studies.

When we look at the vast amount of material that is available regarding patient safety and patient safety culture, a pattern emerges showing that person-centered care leads to better patient outcomes. Our health care systems and their employees are getting better at providing person-focused care, and even with that, they have some challenges. Most have not yet achieved true person-centered care. The difference, while somewhat complex in practice, boils down to the difference of one word: care that is *with* the patient and not just *for* the patient. Most of us have been or will be patients at least at some time in our lives and want our preferences and our perspectives to be respected in our health care. And because we are all individuals, there is no one preference that is right for all of us. To achieve individual person-centered care, individual values and concerns must be considered and respected.

To bring about this change in hospital culture, people from within and from outside of health care will have to join forces and work together. Patients must become meaningful participants in all the conversations that are taking place regarding health care delivery from the exam room to the board room. This is a complex and significant change from business as usual. It will take leaders within all levels of health care. It will require participants from aspiring students to interns, residents, housekeeping, and the C-suites, and from within groups of patients to work together, to bring about small and big changes, before we see the real changes in the culture of health care and patient safety that need to occur.

It is time to stop doing things the same way they have been done for years and take advantage of one of the most underutilized resources in health care: the patient. Many of the systems that have been more successful in embracing this culture of patient safety have embraced patient/person-centered care as well. It has been shown that when patient care is done *with* patients, through "informed choices, safe medication use, infection control initiatives, observing care processes, reporting complications, and practicing self-management" (Rep. Agency for Healthcare Research and Quality, 2014), the results are visible in quality and safety improvement.

As we look back at the blog posts in this chapter, each of the themes, when looked at through the patient perspective, will advance a culture of patient safety. Cassidy Dahn spoke about different ways to think about providing health care to patients. Each of her suggestions would positively answer the question, "What would a patient want?" and each one would also promote a culture of patient safety.

Colleen Parrish wrote about the importance of how language is used in health care pointing out that in and of itself, language reflects and creates the mind-set that is behind the culture. She points out the importance of one's choice of words; the difference between where a word is placed in a sentence impacts whether it is a person-centered thought or not. Word choices change the way people look at situations. By incorporating the patient perspective into the use of person-first language, and through incorporating mindfulness into daily routines, the culture of health care and patient safety is changed. However, for all of this to take place, Garrett Coyan points out that the cultural change needs to occur in the hospitals *and* in our educational programs. He went on to share how little time is focused on patient safety and interdisciplinary communication in his medical school classroom. For the patient experience to be well coordinated, each of the different providers must be able to effectively communicate with each other, a skill that does not always come naturally, but it can and should be taught in medical schools.

I recently read an article by Gurpreet Dhaliwal in which he spoke about how the CEO of a major industrial giant turned around his company in an unexpected way. Rather than changing practices and policies in a number of the business's operations all at once, he focused on the one thing that he felt anchored everything else: worker safety. By placing that one *anchoring* component as the lens through which everything else was seen, he turned the company around. Thirteen years later, its market value had increased by $27 billion, and its safety record was significantly improved. The company was Alcoa. What Dhaliwal suggests is that, in health care, that anchoring concept should be "taking the patient's perspective" (Dhaliwal, 2016). By looking at all things from the perspective of patients and through their lens, he believes that the necessary redesign of health care—the culture change in health care—will be realized.

CHAPTER 13

Why Hospitals Should Fly

John Nance

Blog Posts Written by Health Care Learners

Why Hospitals Should Fly
Eva Luo
June 19, 2012

Despite sporadic episodes of safe, effective, patient-centered, efficient, timely, and equal care throughout my third year, our inconsistent ability to deliver high-quality care has left me almost hopeless about the future of health care. What has reenergized my spirits was reading the book *Why Hospitals Should Fly* written by John Nance, a professional pilot and lawyer with a distinguished career in leading the patient safety movement. The book is a fictional narrative that follows a former CEO of a hospital, Dr. Will Jenkins, as he travels to a suburb of Denver, Colorado, to visit the fictional St. Michael's Memorial Hospital. St. Michael's is *the* ideal hospital that exudes quality, not only in its basic processes and operations, but also in its culture. As Dr. Jenkins visits various departments in the hospital, the reader learns about the effectiveness of specific interventions to improve safety (multidisciplinary rounds, team huddles, checklists, etc.)

and indirectly gains insight into the process of implementation (probably the most difficult part of patient safety work).

When I finished reading the book, I felt like my head had been lifted up from the chaos of our current broken system. My head is now ten thousand feet above sea level, the same elevation where aircraft passengers can safely use their electronic devices. While I'm forced to drink liters of water a day to ward off acute mountain sickness, perhaps it is necessary for me to be at the level where airplanes fly in order to better understand how to redesign our health care system to achieve high-quality care. That is probably the reason why we are all here in Telluride, Colorado.

Here's to a strong takeoff tomorrow!

Another participant blogged about the semantic gamesmanship doctors and nurses and even researchers use to hedge their bets whenever the concept of zero harm is discussed as a realistic goal. Instead, this young doctor took the view that all harm, at least to some extent, is preventable, and that universally we should approach virtually every aspect of practice with proven, highly effective methods for minimizing the risk of harm to near zero. With such an overarching assumption, harm that still results can be attacked from the point of view that we didn't do enough instead of the tired excuse that some harm is unavoidable. In fact, regardless of who may argue that zero is never completely attainable, we have to aim for zero and believe we can get there.

That shift in thinking alone is worth the price of the program: to instill hope.

Initial Reflections

Hope. It is, we're told, a premotion that springs eternal in human experience when the status quo is insufficient or unacceptable. The word and the concept of *hope* is also the essential enemy of what has long been the most dangerous phrase in medicine: "This is the way we've always done it."

Medical practice a century ago seemed fairly straightforward (and totally centered on the individual physician), but it evolved into today's pseudo-system of mind-numbing complexity in which a chaotic patchwork of rules, regulations, procedures, and personal preferences masquerade as true structure. The impact on the people who "live" in that thicket (as well as those who visit) has become a profoundly confusing and oppressive culture that is nearly invisible from within. Indeed, the lethal impact on physicians and student doctors alike encompasses 400 physician suicides and 150 medical student suicides annually (Kalaichandran, 2016). For medical professionals acculturated to knuckling under and living with hundred-hour workweeks, insufficient resources, and a traditional hierarchical framework that suppresses critical communication, the seemingly unchangeable nature of such a stressful culture hides both its dysfunctionality and the fact that it's an inherent threat to patient safety.

New recruits, however, (such as medical students), do see the problems that many within can no longer discern. To the new doctor, the excitement and awe of transitioning from the theoretical world of academics and microbiology to the "real" world of caring for patients is instantly challenged by the cold recognition that they are entering an arena of chaos and intimidation. Yet, their well-intentioned guides—their professors and mentors and those who live professionally within—have neither the patience nor (usually) the temperament to "waste" valuable training time listening to complaints about characteristics they believe can never be changed. So, day by day, the process of acculturation becomes a creeping acceptance.

For example, when she joined TTE as a third-year medical student, Eva Luo (2012) provided the perfect expression of this when she posted a biting blog post that summarized the crux of what TTE is struggling to instill one class at a time: The reality that the status quo is *not* acceptable. It *can* and *must* be changed, especially when it comes to the most basic of all duties of medical professionals to do no harm and to give the best care medical science can provide.

Specifically, TTE strives to do two basic and critical things: permanently open (and keep open) the eyes of younger physicians and nurses to the reality that the present nonsystem of health care is dangerous to patients and inherently incapable of delivering the best care that medical science can provide and to instill a clear mental construct of a safe, quality system. That, in fact, was also the basic goal of "Why Hospitals Should Fly" when it was published in 2009 (Nance, 2009). It provided a shining citadel on the hill, a paradigm of what we can achieve in health care for patient safety and overall quality, based on the seemingly radical belief that we can get to zero harm.

The reason for adopting that same title for this chapter ("Why Hospitals Should Fly") is not that aviation's experience provides all the answers for health care but that the truths we have learned in aviation and many other sharp-end professions about how to keep a human system full of messy, emotional humans safe are universally applicable to medicine. In fact, creating a high-reliability organization requires an end to the concept of the infallible lone eagle. It requires a shared teamwork model that is focused on a shared goal. In the case of commercial aviation, that shared goal is to get you to your destination safely. In medicine, it's doing the best for the patient that medical science, as well as your own education and experience, can provide.

As we discovered in commercial aviation, failing to have faith that we could reach zero accidents would doom us to never getting close. To the extent that aviation could be perfectly safe, which we have at least in part achieved by no major airline accidents in the United States from 2001–2012, so can health care. The problem is, as our blogging student implied, our health care leaders have far too long generally discounted the possibility of actually getting to zero.

Zero harm. In that phrase, there lurks a serious challenge. In fact, as she said, the concept of "getting to zero" has traditionally drowned in a deep and fast-flowing river of cynicism. How (the standard narrative asks) can such a complex system dealing with such a staggeringly complex subject (humans) ever realistically hope to eliminate all harm?

The answer, of course, is that steady, incremental adoption of the methods, procedures, and tactics to shield patients from error can eventually get us to zero. It is true that early successes such as adoption of highly successful standard "bundles" of procedures to prevent central-line infections and ventilator-associated pneumonia are just drops in the bucket of health care's overall performance, but the point is that—however finite in scope—such methods can (and do) drive infection rates to zero when consistently applied. Where one zero can be achieved, others can and must follow. As another student put in his blog "What is Preventable?":

> Our goal has to be zero, perfection; nothing else is justifiable, and yes, it is that simple. Is it pointless to set an impossible, unrealistic goal? No, because there is a point, at least to this one. The fact that our goal is impossible is another way of saying that we should never stop working to be better.

Health care has not so much been traditionally blind to the deadly serious need for improved patient safety as it has been systemically ineffective in knowing what to change and how to change it. The 1999 Institute of Medicine Report, "To Err is Human," had a profound psychological impact profession-wide that has essentially changed the entire mind-set of medicine toward the previous glib assumption that, other than the occasional tragedy, patient safety was not a problem and was adequately addressed by the "quality committees" of hospital boards.

In the aftermath of the IOM bombshell, however, there was little to no understanding of how profoundly the traditional methods of health care served as the drivers of unsafe practices. Hierarchical command structures with physicians trained to be their own authority, for instance, absolutely drove an atmosphere in which subordinates were either fearful of, or systemically reluctant to, speak up when they thought they perceived a problem. Teamwork, as well, encompassed an archaic definition in which the leader barked orders and the followers followed blindly rather than a collegial, interactive team in which the group goal was doing the best for the patient. Even the advent of carefully constructed presurgical checklists and "time-outs" were roundly rejected as being effectively insulting to the practitioners involved rather than understood as new and superior tools for seasoned professionals to use in providing superior care.

In effect, pre-IOM medicine was a physician-centered system in which one human brain was trusted to know all, see all, and perform with unerring accuracy—an expectation we now understand as a matter of certainty to drive the antithesis of a safety system. That assumption is the basic cause of the slaughter of 440,000 patients per year in American hospitals and injury to more than four million others (James, 2013,122–128).

Dr. Don Berwick famously stated, "Any human institution that is built on the expectation of continuous perfect human performance has hard-wired failure into its very structure." Accepting that fact means that every professional must be retrained to admit that—regardless of effort or intent—he or she can make no legitimate promise to permanently refrain from errors. Moreover, understanding the inevitability of human mistakes also requires understanding the essential difference between professional errors and human errors. Professional errors can be controlled and are decisional in nature, and it is reasonable to believe a dedicated doctor or nurse who declares that they will work diligently until retirement will never make a professional discretionary mistake. But human errors largely arise from the inescapable fact that humans do not perceive everything correctly. Occasionally, despite best efforts and a stellar record, even the best and the brightest will make a mistake that, if not caught, could metastasize into a serious or fatal impact on a patient. Tragedies occur when there is no one there ready, willing, able, and trained to have no fear in speaking up to provide the critical "catch." And it is such tragedies and near tragedies witnessed by medical students that contribute to cynicism, despair, and 150 student suicides per year (Kalaichandran, 2016).

What can a dedicated professional do when he or she discovers the inherent inability to guarantee error-free performance? Too often, the response is to simply try harder. But the answer lies in a two-stage process: working hard to minimize human and systemic errors while fully expecting and being ready to intercept the errors that will occur anyway. But the key—the absolute beating heart of patient safety (and, once these principles are revealed, an ethical mandate)—is the reality that only a collegial interactive team (by whatever name) can be effectively trained to expect and catch a so-called error trajectory that may have started with a single mistake. In other words, we will still make mistakes despite our best systemic efforts not to, but the ultimate prophylaxis is the dedication of the team to expect errors and stand ready to catch them in time.

And here, in truth, is one of the key challenges to which TTE is directed: the challenge to teach physicians and nurses how to build, lead, and nurture collegial teams in which each member—regardless of rank or experience—is equally valued, and in which no member would ever hesitate for a nanosecond to speak up if he or she saw, heard, felt, or otherwise perceived a potential threat to the welfare of the patient. Contrast this to the reality of health care's traditional methods in which speaking up is far too often considered dangerous to one's career and welfare.

A 2012 patient safety culture survey reported that 37 percent of frontline clinical staff would not report an error as it was happening, and nearly half of six hundred thousand respondents said that their mistakes are held against them (Sorra, Famolaro, & Dyer, 2012). Collegial interactive teams by any name are not yet contemplated or taught in medical or nursing schools, and that is a major threat to patient safety. Constructing and leading such a team is not a native talent to most, and it requires a certain level of humility and realism that is found in only a few curricula. The very act of building such a team demands a reversal of the "captain of the ship" attitude (or in pop culture terms, the "Captain Kirk" model) physicians have been affirmatively taught and an end to the false assumption that perpetual performance perfection is possible.

While an immediate and profession-wide revolution to this teamwork-driven safety approach is desperately needed, the reality is that the culture will continue to resist the necessary change in philosophy and tactics for the near future. The most effective method in the meantime is precisely what TTE has been designed to accomplish: the medical equivalent of guerilla war. The approach is to teach a handful of new and still-malleable minds to infuse these desperately needed realities in a way that, over time, will grow from within to critical mass and eventually become the norm.

Merely presenting exceptionally bright people with the facts, however, is not enough to counter the momentum of traditional training. The level of acceptance and dedication needed to win over student doctors, nurses, and newly minted residents alike to a true movement requires something more. It requires coming face-to-face with the reality of clinical failure. It requires a direct assault on the concept that doctors and nurses are most effective when the clinical distance between practitioner and patient is profound.

This is not just a matter of asking someone to struggle to maintain a touchstone with the humanity they are dedicated to serve. It is a matter of getting momentarily lost in the excruciating and unspeakable pain, for instance, of a mother's loss of a child when there is no inevitable clinical cause to hide behind—a child, in the case of Josie King, who died in one of America's most renowned hospitals from simple dehydration while all the highly trained nurses and doctors simply ignored a mother's pleas for them to just listen to her observations of her own child. In such stories, we see the operation of a profession that has erroneously convinced itself that the traditional methods of communication and interpersonal relationships are sufficient to provide "excellent" care—and how such assumptions literally kill and devastate. We don't see bad doctors and nurses and techs for the most part; we see humans caught up in the mythology that medicine is a functional system. The reality is that, across most of the planet and certainly the United States, medical practice and what we largely refer to as health care is a great nonsystem still slavishly dedicated to the idea that a physician is omnipotent and infallible because that's what we have traditionally come to expect and what doctors have been taught.

While in TTE, watching the faces of bright, motivated, dedicated young professionals reflecting pain and profound embarrassment is sobering. Watching their faces as the excruciating story of the devastating mistakes that made a near-vegetable of and ultimately killed a bright young man named Michael Skolnik, you also see their faith in the efficacy of the traditional world of medicine shaken, and that is a necessary catharsis. Watching the wrenching litany as Michael Skolnik's anguished parents, Patty and David, agonize over every decision they made on the basis of a questionable surgeon's assurances and recommendations, the very oxygen is sucked out of the classroom by the shared fury at what could easily be the unforgivable negligence of one doctor and his heinous conduct.

However, the lesson that holds the most profound hope that their son did not die in vain is David and Patty's determination that every young professional in the room realize it is their individual responsibility throughout their careers to prevent similar nightmares. The traditional nonsystem was the engine of Michael's demise, a chaotic construct of inordinate faith placed in the nonsensical assumption of infallibility, a nonsystem devoid of the ability to question and catch a major series of mistakes before they could incapacitate and kill. It's also not lost on the room that American medicine is still devoid of an effective method of quickly ridding the profession—and the overly trusting public—of such doctors.

Lessons Learned

Despite nearly 175 years in the Western world of efforts to dehumanize human institutions as the way to protect us from ourselves and our inherent flaws, the reality is that it is always a fool's errand. Embracing our human nature as an integral part of any human institution brings a concurrent responsibility to fully understand how humans fall short of the performance level we desire, and how using that knowledge to build a system in which failures that can't be prevented can be denied the chance to impact a patient. As Dr. David Nash, dean of the Jefferson School of Population Health in Philadelphia, has said, "Care is never error free, but we can make it harm free."

The handmaiden of this methodology and philosophy is fear-free communication, which in a fear-based profession is a tall order that requires nothing short of a massive cultural shift and the eradication of clinical hierarchies. The concurrent reality that just because a doctor or nurse speaks does not guarantee that the message has been received and understood is the mere point of entry for the needed renaissance in embracing how vulnerable we are to communication breakdowns, whether verbal, computer-based, written, or nonverbal. At least 12.5 percent of the time, people who otherwise are identical in training and education fail to understand each other. Under the pressures of daily practice, including fatigue, distraction, linguistic differences, and more, the very assumption that critical information has been passed and understood is extremely dangerous. Focusing on this reality alone can erase nearly 30 percent of unnecessary patient deaths.

It is up to our new entrants and the established participants alike to drop the facades of the past that natter and pay homage to the myth of human infallibility. There is no substitute for understanding that the ultimate prophylaxis against patient harm is a collegial interactive team led by an enlightened person who understands that his or her prowess as a leader is determined by how well that leader can extract, orchestrate, and apply all the human talent entrusted to that leader.

Leaders must proudly say, "I'm very good at what I do, but my strength is in knowing my human weaknesses and providing a highly motivated, finely tuned team around me that is dedicated to expecting errors and never permitting them to hit a patient. That, in fact, is my finest accomplishment."

The profound lessons of TTE cannot be stuffed back in the box any more than a bell can be unrung. The keys to massive improvement and zero harm, once exposed, cannot be ignored with violating an ethical mandate, and those precious and irreplaceable lives that hang in the balance absolutely must be our new North Star.

CHAPTER 14

Mindfulness: It's in the Details

Carole Hemmelgarn and Robert Galbraith

Blog Posts Written by Health Care Learners

Metaphors and Mindfulness
Kristin Morrison
June 26, 2012

I'd have to say that one of my unexpectedly favorite parts of TTE was the absolutely stunning flowers there—from window boxes to hanging baskets to the vibrant fuchsia peonies that currently grace the home screen of my iPhone. I certainly enjoyed them. My week in TTE gave me ample opportunity to stop and smell the roses! But this is a blog post about patient safety and quality improvement, so how is that related to flowers?

On the last morning in TTE, Carole Hemmelgarn asked the learners this question: "You've walked up and down that flight of stairs many times since getting here. Can anyone tell me what's sitting on the windowsill?"

A planter of pink geraniums! I thought. *I notice them every morning!*
Someone quietly mumbled, "Plants?"

I said, "Yeah, pink ones!" because I didn't want to look too eager to share my knowledge of the pink geraniums gracing the hallway of a middle school. Who notices that stuff anyway? Apparently, I do.

But here, pink geraniums are more than a hearty plant that improves the view (like it needs improving). As a group, we reflected about how pink geraniums are a metaphor for mindfulness—that is, the idea that one is fully in the moment and able to appreciate that moment's unique experience. By truly immersing yourself in your environment and taking the time to appreciate your surroundings, you can notice previously missed elements such as a planter of pink geraniums on a windowsill. Additionally, I think there are many varieties of pink geraniums; I can tell you the geraniums are pink, but I can't tell you how many steps there are or whether the nice guy who prepared our delicious food has ever changed his shirt. It seems that working together and generously sharing information would be the best way to get the complete picture about the experience, right?

I think mindfulness is a critically important aspect of patient safety and quality care because it gives health care providers, patients, and loved ones the ability to begin to understand the whole picture of one's health. Paying attention to the small (but important) details provides critical insight into a patient's health and overall well-being, two elements I believe doctors wanting to provide good, patient-centered care keep in mind. I imagine having multiple perspectives—such as those offered by students, residents, nurses, pharmacists, therapists, family members, primary care docs, attendings, and the patients themselves—would help complete the picture. I think patients owe it to themselves to be mindful of their health; after all, they are the ones who ultimately must deal with the decisions made by the team. It seems intuitive that a mindful team, such as the aforementioned one, would be less prone to making medical errors, overlooking important aspects of care, and/or allowing a patient to slip through the cracks. It seems like all health care providers want to provide the best possible care for their patients, and it makes sense to me that part of doing so would involve being mindful of the patient.

I just finished my first year of medical school, which means I haven't had the pleasure of completing any clerkships yet. I'm currently getting some experience working with an internal medicine practice in Saint

Louis, and believe it or not, today was my first day (I think it went really well!).

I'd like to tell you about a pink geranium I noticed today. The patient came into clinic this morning for a follow-up appointment regarding his COPD and diabetes, among other things. My job was to speak with him about how he's been since his last appointment so that I could present him to the doctor I'm working with. We talked about both conditions sufficiently, and while we were sitting there, I noticed that he kept fidgeting with his hands. Unsure how to bring it up with him, I decided to kindly ask him if there was anything else he'd like to talk about this morning. He thought for a moment and said, "Well, now that you mention it, my hand has been bothering me lately." He proceeded to tell me about the discomfort he'd been experiencing in his hand for about the past year (which had gotten worse recently). When he was finished, I thanked him for his time and excused myself to go talk to the doctor.

We discussed my conversation with the patient, and at the end, I brought up the hand discomfort he'd been dealing with. I related that it wasn't bad enough to cause him not to use that hand, but it was bothersome enough that it was often quite distracting.

The doc looked at me, a little perplexed, and smiled. "You know, I've never noticed that, and he's been my patient for quite some time. In fact, he was my dad's patient before he was mine! Good job."

We went back into the patient's room and discussed his COPD and diabetes before examining his hands. After all was said and done, there wasn't much we could do for his hands beyond suggesting some exercises and over-the-counter medications.

But still, I was pleased I had noticed this patient's little quirk and found a way to ask him about it. I'm glad he took a minute or two to speak with me about it, and I am happy the doctor thought it was a valuable piece of information. I doubt it made a big difference in his life or in his overall care, but I think it's a fitting example to show how noticing those pink geraniums can make working with patients a little more enjoyable. I'd like to think that taking the time to discuss it with him helped him realize that his doc (and med student!) really care about the little things in his life.

Back to the pink geranium. As we were walking through her gallery one afternoon, I finally asked her about a painting I had my eye on (she tends to gift me any painting I mention, so I'm careful not to mention too many!). She proceeded to tell me a wonderful story about how that painting was the first one my grandparents sold to her and her husband. She's had it for forty years, and she hoped it ended up in a home that really appreciated it.

A few weeks later, I asked her if it was for sale. My mom's birthday was coming up, and I thought it would be sweet for her to have the piece that her mom sold my new grandma. I knew it would be a treasured addition to our home. True to form, she gave me the painting, and just as I imagined, it's currently brightening the big wall in our kitchen.

So, there you have it: a few geraniums from my recent life.

Cleaning Up Puddles and Addressing the Workaround
Ann Harrington
July 29, 2015

Our morning safety moment consisted of Dr. Mayer speaking about mindfulness versus mindfulness and action. In a grocery store, Dr. Mayer noticed a puddle. He pointed out the puddle to his good friend, Cliff, and continued walking. Meanwhile, Cliff noted the puddle, found a manager, and ensured the issue was addressed and resolved so that no adverse outcomes would result. This safety moment reminded me of the conversation I had with my small group during the domino game on Monday. We spoke of the danger of workarounds, which, in the hospital setting, is the equivalent of noticing a puddle and walking around it.

A workaround I have frequently encountered is the malfunction of the eMAR system at the bedside. Either the computer is running too slow to realistically accommodate the morning med pass, the scanner is broken, or the bar code on the medication itself will not scan.

When these issues arise, the rush of the morning medication administration, patient assessment, and rounds, often necessitate having to go find a portable scanning station to bring into the room. Should I call IT to resolve these technical difficulties? Should I call pharmacy to reprint the med label? Yes to both. Do I? Oftentimes, no. I feel as though I don't have the time.

Why is walking around the puddle an issue? Patients could potentially receive the wrong dose, e.g. Lopressor #50 vs Lopressor #25, the computer prompting me to halve the pill, or the patient receiving the wrong medication altogether. Thankfully, I have not come into contact with such an issue, and when I find a lull in the day, I will take steps to ensure the underlying issue is resolved.

Our Tuesday morning safety moment brought up a common issue many of us have faced in the clinical setting. How do we decrease the incidence of workarounds? What workarounds do my fellow nurses and doctors encounter most frequently in their respective clinical settings?

Wow
Michael Parker
July 11, 2015

I've just completed the second day of TTE, and I can only sum it up in one word: Wow.

The past two days have been both emotionally and mentally draining—hearing firsthand stories from both the families of victims of medical errors and those who are at the frontline in combating them.

The people here have drawn out the battle lines to turn the tide on medical errors and are fighting it with an incredibly powerful tool: education.

Throughout the past two days, I have been privileged to learn about the changing philosophies on optimizing patient safety from families, mentors, and colleagues. I have frequently reflected upon my own practice, especially regarding premature closure and a lack of mindfulness. I have had my eyes opened as to how my attitude toward patient safety can have a huge impact on outcomes for those under my care. Hospitals should be a safe haven for those who are unwell, yet clearly, at present, they are not. Even if that is the case for just one person, it is still completely unacceptable.

I know I still have a long road ahead of me, but I wholeheartedly believe that—through the stewardship of the TTE faculty and the support of my peers here—I can continue to improve my awareness of patient safety. I know tomorrow will bring further challenges, but through these experiences, we can all grow together.

Initial Reflections

Kristin Morrison (2012) uses a beautiful title for her blog post: "Metaphors and Mindfulness." We also see how Ann Harrington (2015) relays a story shared by Dr. Mayer about cleaning up puddles and finds the link to mindfulness. The blog posts address different aspects of mindfulness. Morrison demonstrates the importance of being aware of your surroundings by taking in the pink geranium on the windowsill and how being present in the moment, one of the key elements of mindfulness, helps her notice the patient's twitching hand. In Harrington's blog post, she addresses the next step in mindfulness: taking action. As David Mayer tells the story of his friend, Dr. Cliff Hughes, Cliff was not only mindful about the puddle, but actually took action to have it cleaned up so no one would slip and encounter harm.

On one of my early TTE experiences, Dr. Bob Galbraith taught me the meaning of mindfulness using the very scenario Morrison writes about: "What is sitting on the windowsill?" I have stolen this lesson from Dr. Galbraith because it is so impactful. Day after day, health care workers walk the same halls, yet they cannot tell you the artwork hanging on the walls or who is sitting at the reception desk. These items may seem trivial, but apply the analogy to patient care. Do we walk in patient rooms and no longer see the patient? Do we ask them questions and not really hear their responses? Are we immune to the emotions and visceral feelings going on in a patient's room? How do we bring mindfulness back into our daily lives?

Being present in the moment in health care is difficult these days because of all the distractions and competing priorities. However, mindfulness is a skill to hone and include in a clinician's tool kit. Some individuals will easily learn to include mindfulness in their daily practice, but others will have to work to develop this behavior. As providers, we often ask patients to change their behaviors so they can improve the quality of their lives. We should turn the mirror on ourselves and reflect— how will being mindful help me both personally and professionally?

As you continue to read about mindfulness, I hope I inspire you to have an a-ha moment where you realize the practice of mindfulness will not add stress or time to your day or create more work for you. Instead, it will provide more joy and meaning in your work, help you connect with colleagues and patients, and most importantly, improve care and keep everyone safer in the complex health care setting.

Mindfulness isn't difficult; we just need to remember to do it.
—Sharon Salzberg

History of Mindfulness

In health care, we tend to steal shamelessly, and there is nothing wrong with this philosophy if we are stealing the right stuff. We have learned important lessons from aviation, nuclear submarines, and power plants about how to become high-reliability organizations and apply these concepts to health care. Mindfulness is no different; we are taking the ancient Buddhist practice and incorporating it into the culture of care. Mindfulness is not exclusive to the provider; it is also a behavior attributable to patients and their caregivers.

Buddhists practice mindfulness by being aware from moment to moment, fostering clear thinking and openheartedness, disengaging from strong attachments to beliefs, thoughts, or emotions, and being nonjudgmental (Ludwig & Kabat-Zinn, 2008). Many of the aforementioned attributes evolve as one becomes aware of their stream of consciousness. Mindfulness does not happen overnight. It is a skill that requires practice, but fortunately, as one works on enhancing their awareness in this theory, it can help mold who one is as an individual and clinician.

Mindfulness is also attributed to promoting well-being, an aspect relevant to care providers, patients, and their caregivers. Psychiatry and psychology have created a number of applications based on mindfulness. Stress and anxiety are two specific areas where mindfulness practices have been employed. Also, these practices teach individuals how to handle strong emotions.

At the forefront of bringing mindfulness into the practice of health care are Drs. Epstein and Bishop. They propose the following behaviors are applicable to mindfulness in medicine: attending in a nonjudgmental way to their own physical and mental processes during ordinary, everyday tasks; critical self-reflection; regulating one's attention focusing on the present; awareness of present events and experiences; paying attention in a particular way; and being in the moment (Bishop et al., 2004; Epstein, 1999). These skills are similar to those of the Buddhist practice.

Medical schools are starting to incorporate mindfulness training into the school curriculum. In a medical education review conducted by Dobkin and Hutchinson (2013), they found fourteen medical schools teaching mindfulness to medical and dental students and residents. Applications to health care seem to fall into two somewhat different, although mutually supportive categories. The first is directly related to patient care, and it emphasizes things such as better engagement with patients and colleagues, being more aware of problems and inconsistencies in diagnosis, delivery and coordination of care (particularly as patients age and develop multiple comorbidities), and being more prepared to change the goals of care as patient's needs and preferences change (e.g. from curative to palliative care). The second category is focused on individual providers and their professional well-being and joy. It includes regular meditation, mindfulness-based stress reduction (MBSR), and approaches to mitigate professional staleness and burnout. The effects of this second category on patient care are indirect—but potentially no less important. Interestingly, other learning and service environments like schools, veterans' centers, and prisons have successfully adopted the MBSR strategy.

Now you have a brief history of mindfulness from the Buddhist perspective and how it applies to health care based on the working definition shared from Bishop and Epstein. The next step is to delve deeper into the current state of mindfulness in medicine.

Few of us ever live in the present. We are forever anticipating
what is to come or remembering what has gone.
—Louis L'Amour

The Need for Mindfulness in Health Care

Patient safety issues permeate the practice of medicine. Even though the Institute of Medicine and other reports have identified and recognized these problems, they continue to exist. There have been attempts to increase patient safety through the introduction of electronic medical records, checklists, standardization of procedures, bundles, team training, and creation of high-reliability organizations. However, the fundamental vulnerability remains: we are human, and no matter how hard we try, we will make mistakes.

One of the greatest contributors to patient safety is the ability to pay adequate attention to what is going on in the moment: mindfulness. Medicine has become highly complex, and providers are constantly busy, pulled in multiple directions, and tasked with doing many things simultaneously. The stress induced is antithetical to the effective triaging, ranking, and multitasking essential for optimal patient care. Mindfulness is often replaced by formulaic or even knee-jerk approaches that, while not necessarily deviating from general standards of care, may be inappropriate for a particular patient. This crowding out of mindfulness is not unique to medicine. It pervades our busy, modern lives and causes problems in many other areas.

Health care settings do not naturally breed an environment for mindfulness. Noise is inexorable and pervasive. Digital monitoring means constant alarms, and communications and smartphone technology lead to a flood of calls and texts. This situation of sensory overload is a fact of life, and the provider must, therefore, actively work at remaining mindful, deliberately building time for this into their schedule, preferably as part of a routine (e.g. while eating or at the end of a shift). The other potential benefit of deliberate mindfulness is that the provider is less likely to ignore "softer" aspects of care (e.g. communication skills, professionalism, and well-being of self and team), while grinding out the necessary "hard" work of care.

While there have been great advances in medicine over the past few decades, there is a tendency to look less at the patient in the clinical setting and try to find the answers about them in medical findings supplied by technology. It is important to emphasize that being mindful, means not just being aware and thoughtful in the moment; it is also being present with the patient.

> If we are not fully ourselves, truly in the present, we miss everything.
> —Thich Nhat Hanh

Lessons Learned

How often have you driven somewhere and had no recollection of the process? Frightening, isn't it? However, events like this occur daily in health care. We often refer to this action by saying we are on autopilot, something which is obviously dangerous for our patients and for us. What is missing is mindfulness, a state of conscious awareness of our surrounding ecosystem using all five physical senses, kinesthetic senses, and the happenings of the mind (Brown, Ryan, & Creswell, 2007). Epstein (1999) suggests, "Mindfulness is integral to the professional competence of physicians." It assists in clinical decision-making and helps reduce medical errors.

So, what are some simple techniques that assist in mindfulness? The first is to just breathe. Before entering the patient's room, take a deep breath in, exhale, and clear your mind. This could be added to the routine of washing one's hands and would take no extra time. A deliberate attempt to improve mindfulness may help to have a gut feeling about what's going on with the patient's emotional state and above the medical facts of the case. Are they responding appropriately to current management? How are they feeling about things: cheerful, sad, or anxious? Being tuned in helps us look at the whole state of the patient without restricting ourselves to the appropriate body part or system.

Mindfulness is also about reflective practice. It helps the provider in clinical care by paying attention to images, sensations, emotions, and interpretations—and not just purely in cognitive thoughts. It is a means of broadening the provider's evaluation beyond a narrow focus on theories, diagnosis, and the science toward the whole patient (Epstein, 1999).

Reflective Questions

- Am I present with my patient?
- Am I aware of the nonverbal cues from my patient?
- Am I actively listening to my patient?
- Am I listening to my gut?
- Does this make sense?
- What am I missing?
- Am I seeing physical changes in my patient?

Mindfulness is about noticing, but the important next step is what to do once we have noticed. Mindfulness often calls us to take action. It can mean cleaning up a spill on the floor so someone doesn't slip and fall or asking someone else with fresh eyes to talk about or look at the patient. Taking action when required is an extension of mindfulness.

Leary and Tate (2007) discuss that one of the enablers of mindfulness is limiting our inner self-chatter. We can all too easily become lost in our own heads, with a running commentary of past experiences, future expectations, current bias, and self-judgments. Tuning this down or out is difficult, short of full meditation. However, learning to throw in a mindful "breath" and to be in the moment on several occasions in a busy day can help eliminate that distracting inner voice.

Mindfulness is about self-awareness, but we need to be cautious about becoming self-focused in the process and remembering it is actually about the patient—and not about us. According to Epstein (1999), who quotes Anais Nin, "We don't see things as they are—we see things as we are." Humans frame their world through perceptions and experiences they will have or have already encountered. Biases are part of our being. We must try to step outside ourselves and be aware of the relevant experiences or biases and how they might impact our self-awareness.

There are direct benefits of mindfulness for patients and for their providers. It has been shown that mindfulness is associated with feelings of well-being, fewer depressive symptoms, positive emotional experiences, and regulation of negative moods through awareness, acceptance, and understanding of one's emotions (Leary & Tate, 2007). Communication also improves with mindfulness because there is openness, the ability to process information, and the acceptance of creativity. Connecting and creating trust with patients is important, and engaging in the mindfulness process can help improve interpersonal relations.

Mindfulness helps in creating a safety culture. Weick and Sutcliffe (2006) have studied the impact of mindfulness on the quality of organizational attention. They have identified the following five processes where organizations should focus their attention:

- preoccupation with failure
- reluctance to simplify
- sensitivity to operations
- commitment to resilience
- deference to expertise

One of the points for organizational mindfulness it to remember, but it is remembering in the present and not the past. The past helps build a frame of reference for a particular experience, but it doesn't mean it will be the same in the present situation. Continually experiencing the current environment in the moment is paramount to organizational mindfulness.

There are many opportunities to practice mindfulness. It can be done directly related to patient care, for the well-being of the care provider, or for the benefit of the organization. Mindfulness should really be the default state of mind, and it should not take any time out of a provider's day. It only requires a commitment to engage mindfulness by breathing, honing the skill of clearing the mind, and being aware as much as possible. As Epstein states, "Mindfulness is a discipline and an attitude of the mind. It requires critical informed curiosity and courage to see the world as it is rather than how one would have it be."

Breath is the finest gift of nature. Be grateful for this wonderful gift.
—Amit Ray

CHAPTER 15

Transparency: The Power of Knowing
Tim McDonald

Michelle's Story

In 2008, Michelle entered our organization to undergo the endoscopic placement of liver-related stent under general anesthesia. Her gastroenterologist scheduled the procedure for noon. For a multitude of reasons, the procedure was delayed past four o'clock. After four, the anesthesia service was only providing emergent or urgent anesthesia care. So, the procedure team and the patient were told they could wait until much later in the evening or proceed with the case under procedural sedation to be provided by the GI team—something that had failed two weeks prior when the gastroenterologist could not sedate her adequately enough to tolerate the placement of the stent.

The patient consented to undergo the stent placement after assurances that she would "not wake up" or be in pain as she had been during the prior attempt. Whilst the team sedated the patient, they placed her on her side, and the lights in the room were turned down as they proceeded with attempts to place the stent. Midway through the procedure, the team recognized they did not have the appropriate-sized stent. The nurse assigned to monitor the patient's vital signs left the room to obtain the proper stent from the storage room. She returned to the procedure room after several minutes.

When the nurse returned to the procedure room, she noted the patient to be "shaking." She turned on the room lights to find a seizing, apneic, and cyanotic patient. The patient developed pulseless electrical activity. A code was called. Many mistakes were made during the resuscitation attempts. Eventually, the code team was able to establish a normal cardiac rhythm with a reasonable blood pressure.

When the nurse manager from the ICU arrived in the procedure room in response to code, he recognized the seriousness of the patient's harm and immediately called the patient safety hotline. The director of safety and risk management (DSRM) and the chief safety and risk officer (CSRO) responded immediately to the hotline call. After initially connecting and communicating her commitment to follow up with the family in the waiting room, the director began to investigate the procedure room right after the team had transported the patient to the ICU. A review of the medical record indicated the care was entirely appropriate throughout the procedure until the moment of the cardiac arrest.

Of note, the DSRM discovered some data strips on the floor and in the waste. These were not part of any medical record and were not labeled, but they seemed to be temporarily associated with this case. She then reached out to the patient's family again, expressed empathy, and promised to investigate and share everything with them.

Over the course of the next three days, it became increasingly clear that Michelle had developed brain death. The family was expecting answers. The safety and risk team had concluded their investigation. The situation demanded decisions. In that moment, words and actions mattered most.

Learner Reflections on Open Communication

At TTE, we asked the learners to reflect on open and honest communication in health care. One learner, Rowan Hurrell, reflected on the emergence of recent organizational attempts to focus on patient-centered care and transparency. He stated:

Patient-centered care is the new, hip phrase that has been used by many hospitals in recent years. It is mentioned in briefings and memos, discussed during meetings, and plastered over walls. It is used to demonstrate how hospitals are changing their culture and mind-set as they venture into a new era of modern medicine. The patient is now empowered to be informed and make decisions. It is getting back to the reason why we all went into medicine: to improve the quality of life for the patients we serve. I am completely in favor of and embrace this new approach.

Of note, though, this learner goes on to express skepticism—worried that organizations may be producing excellent talking points but not following through on their commitments. He says, "However, all too often, I am afraid that this new phrase is little more than a marketing gimmick, just a phrase to give the appearance of cultural change without the intended meaning."

In words that specifically related to the current case we are discussing, he goes on to declare:

One way to ensure that hospitals are truly engaging in patient-centered care is to evaluate an institution's transparency. This refers to transparency in outcomes, transparency in reported incidents, and transparency in disclosure. Only when an institution is open, honest, and transparent can measures be taken to learn from mistakes. Mistakes and errors are, unfortunately, inevitable. They are also teachable moments. Only through honesty can one truly examine the systems in place that led to a certain event, and only if one has a transparent system can one get a full understanding of the variables.

Only through transparency can an institution truly embrace a patient-centered culture. I view a transparent institution as one that pays more than lip service to the phrase *patient-centered culture*. It demonstrates a true willingness and desire to learn from mistakes to hopefully prevent future mishaps. The real question will be, Does transparency lead to cultural change—or does cultural change need to happen to improve transparency?

Rowan seems to be "calling the question" for the safety and risk teams in the case for which we are expecting decisions. He essentially asks whether we will follow the path we took in 2002 or blaze a new path of transparency—open and honest communication in the 2008 situation—that involves very serious patient harm with very serious consequences associated with those decisions.

Kristin Morrison, another TTE learner, shared a patient safety story that goes further to ask why we are not always open and honest in our documentation in the medical record, an appropriate question in the context of Michelle's case. Morrison's reflection begins with an introduction to the patient and his medical problem:

> The patient was an elderly gentleman who had been a longtime patient of the doc I'm working with. He had a heart attack a few years ago and has been seeing a cardiologist since then. He recently began experiencing chest pain on exertion, so he went to the cardiologist for an angiogram and possibly angioplasty. He needed two stents.

A week after his angioplasty, she reviewed the cardiologist's report before interviewing the patient. In her words:

> After reading a summary of the patient's blocked arteries and location of the stents, I was shocked (in a good way!) to see this addition at the end: "Adverse event—catheter pierced a small coronary artery. The patient was stabilized, and appropriate treatment was administered. The patient remained overnight for observation. We will follow-up in clinic in three days."

When she queried her supervising attending about such transparency in the medical record, she was told: "Well, there's no reason to hide it if something went not according to plan. People generally like it if you're honest with them, treat them fairly, and tell them about how you'll avoid this in the future." The learner went on to report that the patient wasn't angry. He didn't file a lawsuit, and he still uses the same cardiologist.

When the learner went to see the patient, she asked him about his heart and how his angioplasty had gone.

He smiled and said, "Oh, it was fine. I have two new stents. There was a minor mishap, but I'm okay now."

Poignantly, she said, "It sure is nice to see the principles I learned at the TTE in action."

Another TTE learner, Suresh Mohan, provides a student reflection on truth and humility in medicine—critical components of open and honest communication in health care:

> After digesting for a few days, I've realized our experience at TTE will stay with me. Discussing my week with peers back home, I was shocked to realize how little they knew (and, thus, cared) about the topic of safety. I received responses of, "Well, I guess every field has its downsides" to "Whoa, I didn't know you were, like, super into that primary care stuff." It reaffirmed my decision to have attended and the value of what we learned.

Simply put, he states, "In my view, a lot of issues in patient safety could be solved through two simple things our parents taught us: honesty and humility."

The learner goes on to describe the ways that honesty is imperative when we've committed a mistake. He states we need to apologize and tell the truth. He observes that honesty maintains and builds trust instead of suspicion, and it can potentially prevent the ongoing "after-harms" that occur with a lack of open and honest communication. He cites that research shows that honesty can maintain effective communication, but it saves hospitals money in the avoidance of litigation as well.

An honest admission, though potentially painful at first, will always incite some level of trust in a listener due to its (transparent) nature. Transparency implies respect for the receiver, a rightful expectation of all patients.

As someone interested in one of the surgical specialties, the learner also reflects on the need for humility within that domain of medicine. He stated his belief:

Teaching surgeons humility instead of blind confidence would make for drastically more safe ORs. The humility to know we are privileged to perform such invasive interventions, and to be able to admit to mistakes is paramount. We need to distinguish skill from ego and find examples of those who are able to divorce the two to be our department chairs and leaders. The chair and culture of the department are so crucial to developing an environment where admitting mistakes is encouraged and the lack of which is in fact reprimanded. This idea is contrary to most surgical climates where we reward the so-called perfect, also known as "those best at hiding their shortcomings." Even if these are the only two words that I remember in five years from this incredible experience, I think it'll be worth it.

When Words and Actions Matter Most: The Response

Inspired by their desire to do the right thing—and empowered by the governance and leadership of the university—the clinical team and members of the Department of Patient Safety approached several members of Michelle's family, including her husband, parents, and sister to communicate the outcome of the event review. The CSRO explained that the team providing sedation and the team responding to her cardiac arrest had made numerous mistakes that had caused her anoxic brain injury. These mistakes included the provision of excessive sedation, the failure to properly monitor her respiratory status during the procedure, the failure to establish proper leadership within the code-response team, and the failure to follow standard life-support protocols for a patient with pulseless electrical activity. The manner by which these mistakes caused her death were described and explained until all family members acknowledged that they understood what had happened. The CSRO also explained all the changes that had been put immediately in place after the cardiac arrest and those that would be sorted out going forward.

Following the communication of inappropriate care, the family expressed outrage, anger, and sadness. Those disclosing listened and answered follow-up questions for the next hour. At one point in the conversation, a family member spoke of the concerns about the financial impact of Michelle's death on the ability for her children to become well educated and to go to good colleges since she was the family's primary source of income. As a prelude to the ultimate financial resolution, the CSRO explained that the family need not worry about future financial concerns since the university was prepared to sit down with them at the appropriate time to discuss ways to support them going forward. Following the communication, one member of the patient safety team stayed in the ICU at all times and was available to the family as they made the decision for the timing of removal of life support.

Once brain death was officially declared, the family decided to pursue the possibility of organ donation. The medical examiner initially stated that organs could not be donated because of the need for him to perform an autopsy to determine cause of death in someone who had otherwise been a healthy outpatient. Once it was explained to the medical examiner, with assistance from Michelle's father and a local judge, that the cause of death had been adequately determined and communicated and that resolution was promised, the medical examiner relented and gave permission for possible organ donation.

At that time, one of the father's close friends, who had been the chief of police in the town where the father was a former mayor, was on dialysis at a nearby hospital and in dire need of a kidney transplant. Hearing that news, the team in the ICU helped Michelle's family through the process of a directed kidney donation to the father's close friend. Seven years later, the kidney is still functioning well.

At Michelle's funeral, which was attended by many persons from the university, the team promised to stay in touch with the family over the next several months. A full financial resolution that provided substantial resources to the husband and children was reached within months—without the need for litigation. No depositions were taken. Contact was made with the family every three to four months.

One year after Michelle's death, the family agreed to meet for lunch with the patient safety team to discuss all the changes that had occurred since her passing. During lunch, Michelle's parents asked how they might become involved in patient safety efforts at the university hospital. At that meeting, the CSRO was able to offer the parents and their daughter a seat on the medical staff review board (MSRB)—the group that reviewed all of the serious safety events at the university hospital—as community members, the conscience of the community.

The family has attended monthly MSRB meetings over the past several years and provided numerous valuable insights and suggestions for improvement during that time.

The process by which the university responded to Michelle's death was initially described as the "Seven Pillars approach to adverse patient events" (McDonald et al., 2010).

Potential Involvement and Impact of Participants and Alums from The Telluride Experience

On September 9, 2009, President Obama announced a federal grant initiative focused on funding approaches for the response to patient harm that would put patient safety first and provide rapid and appropriate resolution when inappropriate health care hurts people. TTE faculty members were recipients of those grants and were able to demonstrate some successes to the dissemination and improvement of the open and honest approach to patient harm approach (Mello et al., 2014). These successes have led to the creation of a CANDOR (Communication and Optimal Resolution) toolkit that provides learning materials for many of the principles taught during TTE (Agency for Healthcare Research and Quality, n. d.).

The principles include: the reporting of near misses, unsafe conditions, and harm events, human factors-based event analysis linked to process redesign, open and honest communication from moment of harm through resolution, supporting care professionals after unexpected outcomes, and using data from events to track the impact of changes and identify other opportunities.

The tools are shared with the learners involved with TTE in a way that will facilitate ongoing dissemination and implementation. The TTE faculty members have a long history of effectively teaching learners about open and honest communication (Gunderson, Smith, Mayer, McDonald, & Centomani, 2009).

Learners' Reflections in the Context of the Principled Response to Michelle's Case

Michelle's case does provide an opportunity to comment on the reflections of the students about open and honest communication and their concerns.

The university's commitment to open and honest communication was not a gimmick and was not just marketing. From 2006 through the end of the study period in 2013, the university engaged in thousands of conversations with patients and families following heath care-related harm. This included hundreds of disclosures of errors followed by rapid remediation and learning (McDonald et al., 2010).

It is important to note the team caring for Michelle agreed with the statement shared:

> There's no reason to hide it if something went not according to plan. People generally like it if you're honest with them, treat them fairly, and tell them about how you'll avoid this in the future.

Morrison's reflection captures the essence of the principled approach to harm as practiced by Michelle's team. Sharing these stories with the learners helps show them a better way than the traditional approach to harm that invariably involved a "wall of silence." The team at the university shattered that wall with their open and honest communication with Michelle's family followed by years of contact and engaging them in the patient safety and process improvement efforts at the university (Shelton, 2011).

The two principles appropriate in health care are truth and humility. These principles were apt to describe the approach to the death of Michelle, which was the result of a collection of mistakes and errors. The learner was quite right in sharing his view that such honestly and humility leads to improved physician-patient partnerships, more trust, and less costly litigation.

This new approach that breaks down the wall of silence is a vast improvement over the traditional "delay, deny, and defend" approach. When the faculty share the information related to reduced litigation and the associated costs with the learners at TTE, they provide the learners with data to persuade the skeptics at their home institutions. When the faculty combines the positive patient stories with the data, they connect the learners' hearts with their heads in a way that allows them to be persuasive advocates for change and for implementing and sustaining the principles of open and honest communication in their own environments—when words and actions matter most.

And finally, Cassie Jia, a Telluride Experience alum, describes the impact of her immersion in to the domain of honesty and humility in her posting of "Beyond: Empowerment to Change" when she felt empowered to advocate for others to help make the necessary change.

> I started this program with a general feeling of wanting to be a part of the change, but I walked away with the knowledge of how to be a part of the change. Personally, I recognized the power of words in changing hearts and minds. During the role-playing scenario, I was very touched when the doctor sincerely acknowledged their faults and said, "I am sorry" to the parents. Surprisingly, these three words "I am sorry" carried more weight than I realized, and it made me reflect on the times when I should have extended an apology. Too many times, I had been too prideful with my family. It is much easier to blame the circumstances or even them for what happened.

Imagining myself in this conversation with the family during the same scenario, I feel more empowered to say, "I am sorry that your loved one was harmed because of us. We will never understand how you feel, but we will do our best to walk with your family from now on." I'm empowered now because I understand that improving patient safety is about owning up to mistakes and reflecting on ways to eliminate systemic errors. Health professionals are guests in patients' illnesses, and we need to extend to patients and families our vulnerability and honesty just as they trust us with their lives.

Just as I'm changed by the words shared at Telluride, I hope to encourage and share these words with my peers at home. As I pass through the health care system in my training years, I want to thank and get to know the staff who helped make the hospital a safe, accessible, and welcoming place to be. To start changing the culture around patient safety, I know that change will begin with me. And it already has because I learned to say, "I am sorry and thank you for allowing me to take care of you, my dear patient."

CHAPTER 16

Shared Decision-Making
Jan Boller and Joan Lowery

Blog Posts Written by Health Care Learners

Reflections
Maureen Baker
August 1, 2014

As I watched the Michael Skolnik story today, I wept. I wept for what Michael and his family had to endure. I wept for Patty and David. I wept for the promise of a great kid coming to a close. I wept in fear of what if he had been my son.

I related to Patty when she said she was "vulnerable in the moment," and I reflected on my emergency room visits with my own children. The nurse with twenty years of experience defers to the physician who is willing to fix the gash or the broken elbow at any cost. I did not question, I did not counter. I was at the mercy of their presence and expertise. I plead with the anesthesiologist to take care of my most precious gift.

I realize that even with all of my interprofessional education experience, I have been missing a huge component. The patient and family are a huge part of the equation. The patient and family need to be educated as to how to be a patient before they are in a hospital encounter.

The value of an educated consumer cannot be underestimated and may have the power to change the health care conversation ... from "Do as I say" to "What are your thoughts about how to proceed?" It is a power shift, and I wonder how long it will take to gain collective steam.

I am so appreciative of this experience. I have met a phenomenal group of young people and faculty. I hope to keep in touch with you for many, many years to come.

Thank you for a wonderful experience and for affirming my notion of how important this work is and how fabulous people are fighting the good fight.

Let's Have a Chat
Simone Erchov
August 1, 2014

Most of the week, including today, has focused on methods to improve some aspect of communication. However, the breakdown of communication reaches beyond just the interdisciplinary team, hospital staff to administration, and physician to patient. It applies to the messages health care as an industry communicates to the general public too. As any good PSA will tell you, knowledge is power, and if we continue to strive for a patient-centered/empowered industry, we should seek to educate. Those responsible for the education—hint: us—tend to fall short on delivering a palatable message. Instead, the frontlines of patient education are social networks and blogs. More concerning, the majority of those sources obtain their information from other media outlets.

Now, in theory, that doesn't sound like an alarming statement. The media reports on most issues, health being no exception. It is how the information is being communicated and what conclusions that leads the public—your patients—to that poses the problem. Case in point, a radio broadcast this past May discussed how the media inaccurately reports medical and scientific findings on a regular basis. Consensus suggested this practice continues as little regulation and training exists on how to interpret scientific articles and how to report those findings in a manner that will not lead the public to false assumptions about the research claims. Yet, if whiplash could result from the rate at which the press can turn the same drug that cures an ailment into the root cause of another disorder, we'd all wear neck braces. Some champions of the cause have sought to create a more skeptical public. Others have criticized the press for its behavior. Perhaps we have been missing the mark. Why don't we just talk to the public? It is a recipe for mutually beneficial outcomes.

Communication, whether to an individual or the whole of a nation, is a skill that needs improvement in the medical and scientific community.

The public is hungry for knowledge, but it needs the information communicated in a succinct manner and in a common language. Techniques found in 60-Second Science and PechaKucha support this notion and challenge us to improve. Capitalizing on the use of social media—just as we are here and on Twitter—is a great start. Let's reach beyond just our general network of health professionals and interest groups and attempt to educate. We are the best source of information; let's cut out the middleman and speak for ourselves. Whether blog, vlog, email, tweet, Instagram, or podcast, let's challenge ourselves to improve our communication on a universal level.

Shared Decision-Making
Lakshman Swamy
June 11, 2014

Today's discussions revolved around the importance of involving patients and families in care planning. We touched on a number of important topics within this theme: informed consent, outpatient medication choices, code status, goals of care, and much more. I appreciated that today's SDM discussion was paired with a fantastic workshop on negotiation led by Paul Levy, which broadened the concept of negotiation for me.

Some of the stories regarding SDM were striking. Parents having procedures performed on their children without their knowledge or consent. Patients consenting while heavily sedated. Many terrifying stories today. I think the toughest barriers to involving patients and families in care planning are the incredible asymmetry of knowledge and the difference in perception of severity between provider and patient. What I consider a routine procedure or a common medication may be the biggest risk/benefit decision the patient has ever had to make. We need educational systems to emphasize and formalize our approaches to these questions so that, at the very least, we are clearly aware of when our decision-making is clearly *not* shared and deserves to be.

As Don Berwick shares, "Nothing about me without me."

Initial Reflections

Through the video presentations of real-life tragic stories, TTE attendees are able to powerfully reconnect with what is too often forgotten or overlooked: putting the *humanity* of patients and their families front and center in health care. Medical harm led to the untimely and unnecessary deaths of Lewis Blackman and Michael Skolnik, both bright young men who had their whole lives ahead of them. Following the videos, a parent was present to discuss the situation from the family's perspective and engage with the students in a discussion about why medical harm is so prevalent in today's health care culture.

The experience of bearing witness to these tragic stories, combined with an opportunity to reflect together on the systemic and cultural causes of these types of scenarios, had a transformative and potentially long-lasting impact on the young physician and nurse attendees. Their thoughtful blog posts emphasized the urgent need for health care providers to redesign aspects of medical care and delivery to put the *caring* back in health care. Their concerns focused on how to:

- include patients and their families as fully participating members of the health care team
- shift to a paradigm of patient-and family-centered care that involves doing *with* versus doing *for*
- improve the ability of health care providers to communicate clearly, empathically, respectfully, and collaboratively with all members of the health care team
- ensure that health care providers put patients and their families first and demonstrate this value by responding quickly and comprehensively to medical crises

- make shared decision-making an integral part of any choice points in the patient's health care treatment
- train medical professionals in skills that enable them to reach out to fulfill their roles as educators with patients, families, and the larger community

The above goals will first and foremost take the will and sincere and long-term commitment on the part of the health care industry to provide the requisite training and systemic and policy changes necessary to make a patient-and-family-centered collaborative model a reality.

TTE students' thoughtful blog posts have inspired us to more deeply explore what it will take to actualize the patient-and-family-centered model in the United States. We'll examine the history that led to the current state of health care in this country. We'll call attention to significant breakthroughs that are occurring throughout the industry. We'll examine the chief barriers to family-and-patient-centered care that must be overcome in order to achieve this transformation in health care delivery and provide insight into how these challenges can be met and overcome.

History of Patient-and-Family-Centered Care

For centuries, the reality has been that more than 90 percent of the care given to a person who is ill is provided through self-care or from family and other lay caregivers. Therefore, leaving the person and family out of decision-making during the 10 percent of time that care is currently provided by health professionals makes no sense. Why does it seem to be such a challenge to bring patients and families to the center of health care? French philosopher and historian Michel Foucault provides a plausible explanation in his account of the birth of modern medicine (Foucault, 1963/2003).

According to Foucault, prior to the late nineteenth century, physicians relied on the stories and descriptions of illness provided by the people who were ill and their caregivers. But with medical advances, the physician's "gaze" shifted from focusing on subjective perceptions of the ill person to more objective, technical observations of organs and cells. This was due to improved surgical techniques in the late 1800s, when physicians were able to directly view diseased organs. Advances with the microscope provided the opportunity to look at microbes that cause disease. In other words, the clinician's gaze shifted from what was only subjective (the patient's story) to that which could be objectively verified.

Compounding the distancing between provider and patient as new medical advances emerged, ill people began surviving life-threatening illnesses to such an extent that physicians needed to find ways to care for increasing numbers of patients. They turned to institutions that historically had housed large numbers: military and prison systems. To this day, the "nurses' station" with surrounding patient rooms is modeled after the prison system, where the centralized guard towers could view the inmates. Those housed in these facilities became known as patients, and their families became visitors. Visiting was restricted, and priorities were given to convenience physicians who were faced with increasing numbers of patients. This tradition of hospitals organized for the convenience of health providers has been well established for the past two centuries.

Then, a new problem emerged. With institutionalization, patients faced new hazards of infection, malnutrition, and other life-threatening issues as communicable diseases spread within hospitals from patient to patient. Fortunately, in 1854, Florence Nightingale and her team of thirty-eight volunteer nurses introduced modern professional nursing to health care. Within six months after arriving at the military hospital in Scutari, Nightingale and her nursing team dramatically reduced mortality rates of soldiers injured during the Crimean War from more than 40 percent to less than 3 percent. Nightingale moved the soldiers from dirty straw mats on rat-infested floors to raised cots. She dressed wounds with clean bandages that she had to procure from donations because the military supply officer would not provide supplies to the women. She opened windows for sunlight and invited the soldiers' wives to help feed the wounded from a kitchen that she had built close to the wards. Thus, nurses created the conditions for healing to happen. They worked closely with patients and families to provide a safety net across the care continuum (Dossey, 2010).

However, while the independent actions of Nightingale and her nurses were vital to the survival of the patients, nurses at that time were viewed as subservient handmaidens for the physicians. Their purpose was to comfort the ill. Ignorance about the intelligence, critical-thinking, leadership, and wisdom work of nurses persists into the twenty-first century. Nurses still find ways to do workarounds in unsafe health systems to make health care safe, but, according to ethicist and safety expert Dr. John Banja (2005), these workarounds contribute to persisting safety hazards, which remain a root cause of errors. A team-based, collaborative model—in which health care providers, patients, and families are charged with working together—is emerging with the goals of improving quality, safety, and patient outcomes.

The Evolution of Patient-and-Family-Centered Care

Fortunately, in the second half of the twentieth century, a positive trend emerged, bringing patients and families into hospitals as legitimate care partners (American Academy of Pediatrics, 2012). Research revealed the negative effects of separating hospitalized children and their families, leading to changes in policies around visiting hours. In 1967, the American Academy of Pediatrics introduced the concept of "medical home" in primary care, with an emphasis on family-centered care. Spearheaded by research disseminated by the Maternal and Child Health Bureau of the Health Resources and Service Administration, federal legislation supporting family-centered care emerged in the 1980s and 1990s.

In the late 1980s, advocates for people infected with HIV and AIDS further propelled the movement for patient-and-family-centered care forward. In 1992, the Institute for Family-Centered Care was founded, advocating for partnerships with patients and families across all health care settings. The National Patient Safety Foundation was established in 1997 with a vision of "creating a world where patients and those who care for them are free of harm."

Following the 1999 publication of "To Err is Human," subsequent recommendations from the Institute of Medicine, in the 2001 report, "Crossing the Quality Chasm," identified patient-and-family-centered care as one of the five essential quality and safety competencies to be emphasized across all health professions' education.

The Institute for Patient-and-Family-Centered Care (n. d.) defines patient-and-family-centered care as:

> An approach to the planning, delivery, and evaluation of health care that is grounded in mutually beneficial partnerships among health care providers, patients, and families ... It redefines relationships in health care. It is an approach to care that shapes policies, programs, facility design, and staff day-to-day interactions. It leads to better health outcomes and wiser allocation of resources, and greater patient-family satisfaction.

A Convincing Case for Partnering with Patients and Families

Studies have established the link between patient-and-family-involvement in care with improved patient and family satisfaction, improved patient safety and quality outcomes, improved satisfaction of health professionals and improved cost-effectiveness (AAP, 2012). A systematic national approach has been established to accurately measure and track patient safety outcomes (National Quality Forum, 2016).

In the early 1990s, the Picker/Commonwealth Program for Patient-Centered Care funded a seminal study involving focus groups, telephone interviews, and surveys from patients and families across the United States to gather their experiences and perspectives on health care. This program of study generated eight central themes important to patients and families (Gerteis, Edgman-Levitan, Daley, & Delbanco, 1993; Reid-Ponte, 2008):

- respect for patients' values, preferences, and expressed needs
- coordination and integration of care
- information, communication, and education
- physical comfort
- emotional support and alleviation of fear and anxiety
- involvement of family and friends
- facilitating transition and continuity
- access to reliable care

The Institute for Patient-and-Family-Centered Care (Conway, et al. 2006) identified four core concepts of patient-and-family-centered care (5–6):

- dignity and respect
- information sharing
- participation
- collaboration

Patient-and-family-centered care describes,

> The essential elements necessary to cultivate partnerships at the bedside, in care plan decision-making, healing relationships, and individual patient/professional interactions. More broadly, it provides a framework for formal partnerships with patient and family participation on advisory councils, committees, and design teams (6).

Sufficient standards, guidelines, and evidence exist to make a convincing case that involving patients and families in their care contributes to the "Triple Aim" of better care for individuals, better health for populations, and lower per-capita costs (Institute for Healthcare Improvement, 2016). It is well established that health care works better when patients and families are valued and engaged.

Cause for Optimism: Actualizing a Patient-and-Family Centered Paradigm of Health Care

In the twenty-first century, we have reached an inflection point in the transition from a silo-based, hierarchical, and autocratic model of health care to a new team-based paradigm that includes patients and families at the center of care. Former Institute of Healthcare Improvement (IHI) president and CEO Dr. Don Berwick's statement, "Nothing about me without me" and the compelling mandate, "Do with versus do for" capture the vision of a health care system in which patients and their families are valued members of the health care team. The IHI encourages providers to query patients: "What matters most to you?" as a baseline for creating partnerships with patients and taking into full account what they truly need and want for their care. This is a movement for change that has gathered steam and will not be stopped.

There is much to be optimistic about as we consider the culture shift that is taking place on the frontlines of health care innovation. Throughout this country, pockets of excellence are leading the way toward a more humane and efficient system. Some examples follow outlining how health care industry stakeholders are heeding the urgent call for change on a number of different fronts:

Patient and Family Involvement

Patients and families have been deeply engaged at all levels of health care, helping to bring about institutional and policy changes. Many have experienced the tragedy of medical harm firsthand and are committed to expediting the systemic changes necessary to make health care far safer and more humane.

They serve on patient advisory councils at hospitals and clinics, weighing in on everything from facilities design to system-wide decision-making to pain management and presurgical and postsurgical support of other patients and families. They serve on federal and state commissions where health care policy is being designed. They educate health care students by teaching classes and presenting at conferences. They accompany physicians and residents on grand rounds to contribute the patient's perspective. They serve as teachers to health care professionals to help advance health care transformation, and they are active in many organizations dedicated to promoting a patient-centered model of care.

The Dana Farber Cancer Institute (DFCI) is an example of a hospital with a long-term commitment to instituting patient-and-family-centered care. The tragic deaths of two patients in 2004 catalyzed the need for change. Since then, DFCI has moved far along the path to achieving their goals. They have incorporated active patient and family advisory councils that impact all levels of their health care delivery system.

DFCI's operating premise is to understand the patient as a human being and view the patient's illness through the patient's eyes. No decisions about health care are made without including their patient advisors (Reid Ponte & Peterson, 2008).

DFCI seeks to incorporate the patient perspective in everything they do. This includes:

1. Engaging in cross-functional collaboration, communication, and decision-making involving frontline staff, patients, and their families.
2. Shifting from a culture of blame to a culture of learning. They embrace nonpunitive reporting. They are guided by values of fairness and justice.
3. Identifying errors and near misses, promoting open discussion, sharing problem-solving, and learning.
4. Incorporating systems thinking. They partner and learn from all players on the team.
5. Creating a fair and just culture. This requires giving feedback skillfully and undergoing critical analysis.

Health Care Redesign

The Southcentral Foundation's Nuka System of Care is a leading-edge model of health care redesign with customer-owners. *Nuka* is an Alaska Native word having multiple meanings, including *honor, strength, big living things, dignity, love, generosity,* and *spirit* (Southcentral Foundation, 2016). Its name captures the deeper meaning of the nonprofit, integrated health care system that was created by and for Alaska Native people. It is managed and owned by Alaska Native peoples with the goals of achieving physical, mental, emotional, and spiritual wellness for their community. Based in Anchorage, it services more than sixty-five thousand Alaska Native people in Anchorage, the Matanuska-Susitna Valley, and fifty-five rural villages.

The Nuka System of Care calls itself a *relationship-based system* that includes organizational strategies and processes, medical, behavioral, dental, and traditional practices, and an infrastructure designed to support wellness. A 2011 recipient of the Malcolm Baldridge National Quality Award, Southcentral Foundation's Nuka System of Care is recognized by the National Committee on Quality Assurance as a patient-centered medical home level 3—the highest level. Their system is built upon three central beliefs:

1. Customer ownership. They are not called patients. They are known as customer-owners and are treated as equals in shared decision-making around health and wellness issues.

2. Relationships: The relationship between the primary team and the customer-owner is considered central to effecting change. The team focuses on understanding each customer-owner's values and influences, respects their lifestyle and health care decisions, and supports them on their path to wellness. To benefit from a variety of perspectives, customer-owners meet with the entire team together. Same-day appointments are available, and phones are typically answered within thirty seconds.

3. Whole-system transformation: Customer-ownership and relationships are at the center of the entire system, extending beyond health care delivery. Workforce systems and training support the care model, impacting how human resources hires employees, the setup of administrative and support teams, how employees interact with each other, and facility design and workspace sharing. All systems were designed to foster respectful, team-based, collaborative relationships.

Since its beginnings in 1999, Nuka has experienced a 70 percent decrease in hospital admissions and hospital days and more than a 30 percent decrease in outpatient visits. Customer-owner and staff satisfaction has remained above 90 percent for many years. They are achieving greatly improved medical outcomes while reducing costs. In addition to modern medical services, the Traditional Healing Clinic provides traditional Alaska Native approaches to health. Their goal is to incorporate customer-owner values, beliefs, and practices into the healing proposition.

Cultural Brokers

There is a great need for people who can bridge the cultural divide in health care settings. A cultural broker is someone who can apply cultural knowledge, health science knowledge, and skills to negotiate with the client and the system to achieve beneficial, effective health care planning. They serve as cultural guides, liaisons, mediators, and change agents in collaboration with health care providers (National Center for Cultural Competence, 2004).

The Cultural Broker Project's mission (under the auspices of the National Center for Cultural Competence) is "to increase the capacity of health care and mental health programs to design, implement and evaluate culturally and linguistically competent service delivery systems" (National Center for Cultural Competence, 2004, 1). This is a critical step toward achieving the goals of eliminating racial and ethnic health care disparities and improving health care literacy.

Cultural brokers include a range of people: immigrant children who navigate through two or more cultures on a daily basis to community peer mentors; educators, health care professionals, and staff working at all levels within health care institutions; and organizational leaders. While training and experience vary, they are skilled at navigating between cultures and are knowledgeable about the health values, beliefs, and practices within their cultural group as well as the health care system that services their community.

In the Latino community, the Promotor movement is growing rapidly and leading to improvements in prevention and treatment. A Promotor is a lay Hispanic/Latino community member who receives specialized training to provide basic health education and advocacy to community members (Latino Health Access, Visión y Compromiso, & Esperanza Community Housing Corporation, 2011).

Western University of Health Sciences College of Graduate Nursing (CGN) has developed a model of collaboration with a group of Promotores. Together, they are finding innovative ways to engage residents in the city of Pomona, California, which has a 70 percent Latino/Hispanic population, in working toward achieving the Triple Aim through community, education, and provider partnerships.

The CGN initiated a process to prepare Promotores to bridge gaps in knowledge about health, safety, and health care across Pomona's neighborhoods. Within one year of receiving training about disaster preparedness, safety, preparing healthy foods, and engaging in exercise and active living, Pomona's Health Promotores, working in collaboration with Western University faculty and students, have provided information and experiential education to more than two thousand Pomona residents. Ongoing requests for more Promotor education have exceeded expectations. The Promotores have an office at the college. Partnerships in design and delivery of health and safety education happen on a daily basis and co-learning between partners is rich.

Transforming Health Care Education

As previously discussed, interprofessional training between future health care providers is key to achieving the goals of patient-centered, team-based care. Training health care professionals of tomorrow in emotional intelligence-related skills and attitudes will ensure that they can succeed in an environment where empathy, self-regulation, listening, and communicating effectively will be required to build strong partnerships with colleagues, patients, and their families.

Design Thinking, Health Care Innovation, and Community Outreach

Kaiser Permanente's Garfield Innovation Center in Oakland, California, has adapted principles of design thinking to envision and prototype the "hospital of the future"—a vision that puts patients and families at the heart of care (Garfield Center, 2012). Design thinking is a "people-first" approach that facilitates innovation by seeking to understand what people need and want and then applying the designer's sensibility and methods to create solutions.

The Garfield Center is the largest health care innovation center in the country. Its goal is to "inspire people across all regions of Kaiser Permanente to imagine the future of health care and give them the tools to create it." After homing in on problems that need addressing, their teams engage in hands-on activities such as simulations, technology testing, prototyping, and product evaluations. Successful initiatives are further refined at Kaiser medical centers, offices, and clinics before being rolled out across the nation.

Their work has led to a portfolio of innovations that touch all aspects of care—from facility architecture and interior design to innovations and inventions designed to improve workflow, improve the efficiency and effectiveness of services, enhance the patient experience, create advanced technologies, and even reimagine the conventional doctor's appointment.

These and numerous other examples of how health care is evolving on multiple fronts toward the goal of putting patients and families front and center give us cause for optimism about the future. TTE scholar and medical student Simone Erchov captures the essence of what it will take for each physician to personally model the transformation we are striving for. Her words are relevant for *all* health care practitioners to reflect upon:

> Considering the patient as we would our own family, involving the patient in the treatment process, providing the alternatives and associated risks during informed consent, apologizing, and admitting a mistake—these are all behaviors that acknowledge and enliven the interaction between patient and physician. It builds relationships and trust. It changes the course of outcomes. So, perhaps it is important to reset our mentality and focus on a simple, yet important component to which we all can relate—being human and the needs and desires that condition requires.

Challenges and Lessons for the Future

While much has been achieved, it is clear that in order to fully transition to a patient-and-family-centered paradigm, a sustainable commitment to change must take place on the parts of *all* stakeholders: administrators, the full spectrum of health care-industry workers, payers, planners, policy makers, educational institutions, patients, caregivers, and the community at large. There is abundant evidence that this new model improves quality and safety, is better positioned to achieve desired medical outcomes, and results in increased patient and provider satisfaction, greater efficiency, and reduced costs. However, it is also evident that in order to achieve a long-term patient-and-family-centered model an ongoing and vigorous commitment to sustainability, continuous reflection, cross-cultural understanding, and learning is imperative.

The Road Ahead

> I think the toughest barriers to involving patients and families in care planning are the incredible asymmetry of knowledge and the difference in perception of severity between provider and patient. What I consider a routine procedure or a common medication may be the biggest risk/benefit decision the patient has ever had to make. We need educational systems to emphasize and formalize our approach to these questions so that at the very least we are clearly aware of when our decision-making is clearly not shared and deserves to be. As Don Berwick shares, "Nothing about me without me" (Swamy, 2014).

Challenges must be overcome in order to actualize a patient-and family-centered care model. Already, stakeholders and organizations are meeting these challenges. However, the road ahead will likely be bumpy, and being prepared for the journey and keeping the end-goal in mind can help us get to our destination quickly, safely, and with determination.

Solutions

Baker states:

> The patient and family need to be educated as to how to be a patient before they are in a hospital encounter. The value of an educated consumer cannot be underestimated and may have the power to change the health care

conversation ... from 'Do as I say' to 'What are your
thoughts about how to proceed?' It is a power shift, and
I wonder how long it will take to gain collective steam.

So, what will it take to bring patients and their families back to the center
of health care? Six recommendations are offered, which are consistent with
recent established priorities identified by the Josiah Macy Jr. Foundation
(2014) and the National Patient Safety Foundation (2015).

1: Move from "power over" to "power with"—from "caring for" to "caring
with." First and foremost, a paradigm shift in prevailing attitudes must
take place. Early in the twentieth century, Mary Parker Follett introduced
the idea that a "power with" attitude, must replace the "power over"
mentality within communities of common interest so that collective
wisdom can emerge, unhampered by restrictive hierarchical dominance
(Briskin, Erickson, Ott, & Callanan, 2009).

2: Recognize that health care is a community-wide effort. We must
understand and accept that everyone involved in health care belongs to
the same "community" in which everyone contributes to better outcomes.
Wenger and colleagues have proposed the idea of "communities of
practice," where people with common interests and a wide variety of
expertise work collaboratively to fuel change for improvement. All
members of the community—in the case of health care, that includes
patients, families, health professionals (including students and faculty),
and others who are involved in health care—are considered "legitimate
participants." Patients and families are not visitors, and students and health
care faculties are legitimate participants in care (Wenger, McDermott,
& Snyder, 2002). Health care is not a private club with a password for
entry. Like the Nuka System of Health Care, "patients" are regarded as
"customer-owners" and the raison d'être for health care delivery.

3: "Tear down the walls of the past!" The attitudes, behaviors, and
organizational barriers that have created the hierarchy and dysfunction
in health care need to be dismantled and reinvented if the new model
is to succeed. For a transformation to take place, health care providers

must envision themselves as team members with the patient, not the provider, at the center of care. This will require motivation and training to develop and support communication skills that focus on listening, empathy, nonverbal awareness, cross-cultural sensitivity, and the ability to communicate with clarity, caring, and honesty. It will require self-reflection and developing an increased self-awareness about how one is perceived by others. Ultimately, it will require flexibility and a willingness to be challenged and grow in new and perhaps unexpected ways.

4: Empower and partner with patients and families. Patients have also been acculturated to feel intimidated by their health care providers and not question what they are told. Many have adapted to feeling "lower status" than their physicians and do not feel comfortable questioning or offering their input. Some are simply more comfortable being told what to do versus questioning medical decisions or becoming more proactive as patients. If a transformation is to occur, patients must be educated about what it will take to be partners in their care and understand why they should buy into this new paradigm. They must be encouraged to speak up, and they must be supported in learning ways to function as critical members of their health care team. In short, they must be invited into the conversation in a way that enables them to add value to the treatment process.

5: Update health care policy. Patient-centered care policies must move beyond the designated "health care/medical homes" to all of health care. Health policy reform at all levels is needed to ensure that patients and families are participants in care. Small steps have been made requiring better identification of health providers so that patients and families know who they are working with, such as South Carolina legislation stemming from the Lewis Blackman case. Advocacy from the late Josie King's family has spurred changes in hospital policies to empower patients and families to call for rapid-response teams (Landro, 2009). Funding needs to move beyond limited physician-and-facility payment to reach all entities vital to patient safety. This includes funding community assets to create models of prevention and assistance before patients become too ill to care for themselves.

6: Groom transformational leaders through transformative learning. It will take transformational leadership and transformative learning to help shift perspectives from the old "power over" paradigm (caring for) to the "power with" paradigm of caring with. Transformative learning is "an approach to teaching based on promoting change, where educators challenge learners to critically question and assess the integrity of their deeply held assumptions" (Mezirow & Taylor, 2009). These new perspectives, coupled with experiential learning and collaborative dialogue and practice, lead to far-reaching change.

TTE exemplifies the transformative learning that will ultimately produce the next generation of transformational leaders. This is a style of *leadership*—where the *leader* works with teams to identify the needed change—creates a vision to inspire and guide change and execute the change in tandem with committed members of the group. TTE scholars leave with a commitment to execute quality and safety projects from start to finish in their respective learning communities.

These six recommendations represent a paradigm shift that will take time and patience since it means moving from novice to expert, a process that involves the acquisition of new skills, attitudes, and understandings (Dreyfus & Dreyfus, 2008).

Concluding Thoughts

The journey from novice to expert is a winding road with unexpected turns and fitful stops and starts. As medical student Erchov reminds us, the goal is simple, yet complex, and worthy of our patient, unwavering focus as we seek to actualize the kind of health care we all want for our loved ones.

> Errors happen. We are human, systems do not remedy themselves overnight, and medicine is a high-stress, high-risk endeavor. But involving the patients, empowering them even, allows for the reintroduction of a key element to the doctor-patient interaction. This element is so important yet often overlooked; it is the

element of humanity. The human element demands respect, empathy, communication, and understanding. It seems so intuitive, yet reports on these cases consistently identify such basic human elements to be lacking. Several reasons can be argued as the source of this dearth, but nonetheless, it should be reintroduced as a focus.

CHAPTER 17

Leadership
Richard Corder

Blog Posts Written by Health Care Learners

Process Improvement in Real Time
Julie Morrison
July 10, 2012

Upon returning to UT Medical Center from TTE, I was filled with motivation and a new sense of purpose. Something had changed in me out there. I used to be more of a *thinker* (forever told I would be good as an internist) and not so much of a *doer*. I struggled with this during my third year of medical school because I saw so many areas of improvement, but I often sat near the sidelines gathering information rather than jumping in and stimulating change.

At TTE, I was so encouraged by the perspective of the senior faculty; the fact that they found similar aspects of the clinical world frustrating or inefficient and were looking for a collaborative team to face these challenges reinforced my perspective and encouraged me to take action. The training gave me a better vision of the organization and hierarchy of a hospital as well as the tools to accept such a call to action. I realize now that a vision cannot be translated into change without the support of various members of the care team and hospital administration. A good idea can fester forever within someone without ever seeing the light of day or be implemented without the correct support and flop because of a lack of motivation or proper support.

On the third day of the conference, we broke into small groups and discussed real-life changes we would like to see in our hospitals. My team came up with a "bedside communication whiteboard" that would list the care team (nurse, attending, residents, medical students, PT, OT, RT, etc.), their expected procedures, and the daily care plan. It would also serve as a place for the patient or the family to list questions and concerns. We envisioned that this tool would empower the patient to engage in their health care by making them informed participants and serve as a stopgap from wrong procedures. When I returned to campus, I was excited to stimulate change. However, I tempered this excitement in order to stimulate the proper support for my ideas since I didn't want to be run over by resistance. I scheduled a meeting with the CMO of our hospital system and settled into my clinical rotation.

The first day, I spied portrait-covered papers in several patient rooms before a nurse manager came by and left a stack on the cart we were using for rounds. I immediately picked them up and found a paper form of our bedside communication whiteboard! I almost couldn't contain my excitement. I followed the nurse down the hall and asked her about the innocuous sheet of paper. We engaged in a great conversation, and she invited me to sit in on the quality-improvement meetings she attended.

I took over the responsibility of filling out the sheets for each patient as we rounded for the rest of the week. I did so silently at first. And then, I started getting questions from the other members of our team. In the past, I would have been uncomfortable explaining my actions. I would have been concerned that it wasn't a big enough idea—or that they would find it frivolous or frustrating. But not now; because of TTE and the perspective of the senior faculty, I spoke with confidence and clarity of vision. I knew that this tool would make a difference, and I educated the other students, residents, and even our attending so that they could see its value as well.

By the end of the week, I was receiving really good feedback from the patients. They said, "Oh, I was waiting for this," and "This is so helpful. Thank you." The senior resident even thanked me for utilizing the sheets.

I think it is going to stick, and I think we will have it implemented across the hospital by the winter!

Fight the Fight

Junzi Shi

July 19, 2012

A soldier cannot run from battle because there are guns trained on his back in both directions. This is not so in other occupations. As medical students and professionals, we constantly have to choose how to react to new information and decide whether to take action or sit on the sidelines. Soldiers don't have the luxury of these options. So, in a way, because we must make difficult choices, we also need to muster more courage to perform in the battlefield.

Patient safety improvement takes a lot of guts. It means challenging the status quo, confronting established traditions, risking your job or grades, and most of all, dealing with recalcitrant people … who may be your seniors.

Today, I was confronted about my experience shadowing a preceptor who does not wash his hands before he sees patients. I have commented on this directly to the physician by saying, "I notice that you don't wash your hands between patients. This is different from what we learned in school." This did not change his behavior, but I did notice that he would wash his hands after he saw me do it first. I have been rather passive about confronting my preceptor again, but I think I have renewed determination after these eye-opening conversations at TTE.

Good leaders set priorities and make decisions. TTE is doing more than informing us; it is training us to be leaders. We will not turn back in the face of adversity, we will fight against apathy, we will be soldiers who cannot turn back because we see no choice but to push on.

Power

Maureen Baker

August 1, 2014

As I reflect on a wonderful first day, the notion of power keeps coming up as I ponder how to bring about this much-needed patient safety revolution.

The power structure needs to be shaken up a bit, falling heavily into the hands of the patient and family. No longer should patients be at the mercy of a fragmented and hierarchical system.

How can we redistribute the power? Educate the young, regulate the old, and gain genuine buy-in and support from organizational leadership. Who has the power to regulate the old and influence leadership to embrace a culture of patient safety. How do we convince those in power to give it up for the sake of patient safety?

Initial Reflections

My immediate reflection is that each of the learners attending TTE, contributing to the conversation, and sharing their thoughts, challenges, and lessons learned is a leader in the purest form of what it means to lead. These are stories of their leadership journeys.

The TTE blog post entries all share a clear theme, whether a reflection during the time together or a story upon their return to their clinical environments. The authors share a sense that they are changed for the better. Their thinking has been challenged, and their ideas have been given a voice, and that voice has been validated by fellow students and an engaged and open faculty.

The conversations, exercises, and reflections of the time together in TTE seem to leave participants with a renewed sense of purpose and an opportunity to challenge long-held mental models that have been reinforced and rewarded.

The participants share a belief that there is an existing power structure in health care that is not in the best interests of patients or families. They explore the notion that to shift the current power structure, they are going to have to think differently, be open to seeing and listening to the current state in new and different ways, encourage the guts and bravery necessary to speak up, act differently, and do so in a collaborative, team-based environment. They speak of the need for a patient safety "revolution," a call to action to redistribute the power of the decision, the conversation, and the care to more respectfully and holistically include everyone's voice. Most importantly, the focus is on the voice, needs, and wishes of the patient and their families.

This notion of speaking up appears to be a reflection of each authors' inner voice of sorts. Their blog post entries are akin to accounts of an internal conversation that appears to be born out of a natural sense of self-preservation in the climate to which they have grown accustomed. Certain themes and thoughts are clearly shared through these essays: What will others think? Will I be perceived as stupid or silly? Will my idea or suggestion be met with resistance or ridicule by those more senior than me?

This sense of self-doubt—and the subsequent reluctance to speak up or do anything—is an understandable reaction to the environment that has become their norm. These environments have reinforced models of reward and recognition that meet challenging the status quo with the risk of failing grades, undesirable schedules, or even putting your job at risk.

The authors returned to their clinical work with a sense of optimism and excitement that appears quickly tempered by the realities of the organizations to which they belong. While they have practiced, witnessed, and been encouraged to be the brave voices that speak up in the face of apathy and the current power structure, they return from TTE without the support of their fellow students and supportive faculty. They return to a sense of being a lone, small voice in an environment that is unchanged from when they left.

They speak eloquently of the realization that lofty or ideal visions for a better future do not translate into action or change without the collaborative efforts and conversations of everyone on the care team.

Leadership: A Historical Background

All of the great leaders have had one characteristic in
common: it was the willingness to confront unequivocally
the major anxiety of their people in their time. This,
and not much else, is the essence of leadership.
—John Kenneth Galbraith

The history of leadership in health care mirrors the most widespread theories of leadership since the mid-1800s: *great man theory, trait theory, behavioral theories, contingency theories, transactional theories,* and *transformational theories.*

In a nutshell—and not to make light of the significance of each of these respective areas of study—we have evolved from the idea in the mid-1800s that great leaders were born, not made, and that leadership was intrinsic (and as the great man theory name suggests, only a man could have the characteristics of a great leader). This gave way to the trait and behavioral theories of the mid-twentieth century when we began to understand that with the right "conditioning," anyone could have access to the previously exclusive "birthright" of naturally gifted leaders. In other words, over the past 150 years, we have woken up to the realization that leaders can be developed, nurtured, and trained.

What is of note in health care is that when we reflect back on the relatively young field of our organized approach to hospitals, medical education, and the formalization of health care as an industry, we find safety-minded leaders whose learning, teachings, and approaches have apparently been lost or forgotten over time.

Florence Nightingale (1820–1910) challenged a male-dominated establishment (the British government, the army, and the church) to bring order and hygiene into a chaotic, disease-ridden organization.

Her story is rich with examples of speaking up, using data to support her hypotheses, establishing clear rules and following them, and securing basic supplies while facing continuous opposition to her ideas from the army and from the medics in the field. She shared data (some of the earliest infographics) that a larger number of soldiers were dying from what we would call HAIs today than from wounds on the battlefield.

She expected nurses to wash their hands and insisted on the laundering of linens between patients. She was a patient safety advocate and leader way before her time. Her ability to get the job done—in difficult circumstances while facing strong opposition—shows the leadership values of hard work, following a bold vision, collaboration with others, and the use of data to support our improvement efforts.

"For the secret of the care of the patient is in caring for the patient." These words closed a lecture given by Francis W. Peabody (1881–1927) to Harvard students on October 21, 1925. The lecture that he entitled "The Care of the Patient" deserves reading more so today than perhaps ever before, particularly when our profession, payment models, approaches to education, and medical technologies seem to focus more on the disease than on the patient.

In his short life, Francis Peabody's impact as a physician, researcher, teacher, and person was felt by many. He attended Harvard Medical School and interned at the Massachusetts General Hospital. At Johns Hopkins, he continued his training with the likes of Dr. William S. Thayer. Over time, he grew and was mentored as a clinical investigator, a bedside teacher, and a leader.

These two arguably junior, young, inexperienced, radical leaders made a personal commitment to do their work differently. In essence, we are reaching back to the lessons of the past and remembering that much of what guides us has not changed. The call for safety now is the same as it has always been: "above all, do no harm." And it is not reserved for those with title, conventional authority, or experience.

Dr. Lucian Leape reminded us in his 2008 piece on the scope and problem of patient safety that the present-day "movement" was most likely born out of the publication of the results of the Harvard Medical Practice Study in the *NEJM* in February 1991. While this study was driven by the medical malpractice crisis at the time, the investigators uncovered glaring realities of medical injury and adverse events. The data provided a shocking look at the amount of injury and death related to medical care, and it provided a wake-up call to many with the acknowledgment that much of it was avoidable.

From here, we have heard a regular "drumbeat" of IOM Reports, publicized tragic stories of failure, and wonderful, team-based, data-driven improvement stories.

Governing boards, executive leaders, caregivers of all disciplines, middle managers, and frontline staff of health care-delivery organizations are being increasingly encouraged to think in terms of building high-reliability organizations, leading cultures, and environments that are more inclusive of different opinions, and embracing ideas, approaches, and methodologies that continue to inform us as we learn, develop, and grow as leaders.

Leadership: Current State

> An organization cannot be what its leader is not.
> —Richard Corder

By current estimates, there are more than fifty-five thousand books with the word *leadership* in their titles available on Amazon. In 2015, it was estimated that more than four new books a day were published about leadership, and there is no indication that this is going to let up or change any time soon. And this doesn't include the additional articles, journals, and other efforts to teach, learn, share, and improve in this space.

Nowhere has this growth in the interest of leadership been more apparent than in health care. Perhaps this is due in a large part to the fact that while we have clearly established that leadership can be learned (leaving the great man theory where it belongs—in the mid-1800s), we have also, perhaps a little too slowly, been awoken to the fact that our current approaches and systems are not creating safe environments and processes for the most effective, timely, cost-effective, or safe delivery of care.

The reawakening that we have been going through since the early 1990s in patient safety leadership has us reflecting on the lessons from the past, seeking the expertise, experience and science from other fields, and working to train and educate our young so that we can break the cycle of passively accepting avoidable harm as a necessary evil.

While there are pockets of highly reliable, safe, harm-reducing systems throughout North America, from listening to our TTE faculty and students, these remain pockets. The prevailing norm is more of a tolerance of mediocrity than a drive toward standardization and zero harm. There are still too many stories of situations that could have been prevented and people harmed because of failed systems, tolerated practice, and behavior variation.

We are exposed to stories of people not leading with humility and respect, not listening to the opinions of others, and treating dissenting points of view as disruptions to be ignored or eliminated.

I heard a story this week from a hospital senior manager after having visited her deceased brother-in-law in the hospital where she works. She had driven her sister to the hospital, after her sister had received the news thirty minutes earlier that her sixty-two-year-old husband had died suddenly and wouldn't be coming home. They walked into the room where he was lying, and no one had taken the time to remove the tubes from his mouth or clean up the clearly vigorous efforts to resuscitate him.

My colleague told me through tears that this would be the final memory she and her sister would have of her brother-in-law (her sister's husband) and that it embarrassed her that her hospital didn't care enough to clean him up, restore a little dignity, and treat the deceased with some respect. On reflection, she stated out loud that as a "non-clinician in this academic medical center, saying anything to a colleague or to the nursing staff would not be well received, had not changed practice in the past, and why bother rocking the boat."

If we believe that leadership can be learned, and we agree that we need to do a better job, then we have much to learn. There are many indications that our hospitals and health care system in general have significant opportunities to be better led.

Lessons Learned

Control is not leadership; management is not leadership; leadership is leadership. If you seek to lead, invest at least 50 percent of your time in leading yourself—your own purpose, ethics, principles, motivation, conduct. Invest at least 20 percent leading those with authority over you and 15 percent leading your peers.
—Dee Hock

As I said in my opening reflection, I am of the opinion that each of the learners attending TTE, contributing to the conversation, and sharing their personal thoughts, challenges, and lessons learned is a leader in the purest form of what it means to lead. What is leadership anyway? Let's start with what it is *not* leadership.

Leadership has nothing to do with seniority or position. I hear too much talk about leadership referring to the senior-most executives or clinicians in a hospital. Remember, they are just that: senior executives or senior clinicians. I have seen more authentic leadership from a third-shift oncology nurse, a first-year medical student, and a thirty-year housekeeper than I have from some seasoned senior executives with many years of experience.

Leadership is not about managing people. Management is critical to running a safe hospital. We need to nurture, train, and support managers to plan, measure, coordinate, hire, and fire—and so many other things. Leadership and management are not synonymous.

Leadership is a personal commitment to change the world (however *you* define the world: yourself, your church, your home, your hospital, or your practice) through influence and example that maximizes the efforts of yourself and others around you and achieves the change as you've articulated it.

Steve Farber, author and business leadership expert, says, "Approach the act of leadership as you'd approach an extreme sport: learn to love the fear and exhilaration that naturally comes with the territory. To become an extreme leader, you need to seek opportunities that will stretch you and ultimately cause fear."

This construct and way of thinking resonates for me in the work of leading to improve patient and provider safety because so much of what we hear is about the fear of speaking up, the fear of having a different opinion, the fear associated with disagreeing with the establishment, and fear of suggesting changes to how we provide and support patient care.

If we embrace the fear and act, we become leaders. Our patients deserve nothing less. We can embrace our fear to reduce the fear of those we care for.

Leadership is a personal choice. It can be learned, and it evolves. As leaders, we are forever learning, changing, and adapting. It is worth remembering that leadership is not reserved for those with title, authority, or position within our hospitals. This is of critical importance when reflecting on this issue of the needed shift of power, away from the establishment (the doctors, nurses, and others on the care team) to a more informed and balanced model that includes the patient and the family as an equal partner in care.

Leader, Know Thyself

Taking the time to understand our own motivations and remembering that we are unique individuals, although all called to similar work, is an important part of developing our leadership skill sets and competencies. It is time to abandon the "Pygmalion Project" of the endless, and frankly fruitless, work of trying to make everyone conform to our way of thinking or doing. As leaders, it is important to self-reflect and then frame (write down) what success looks like for us.

- Self-Assessment. A wide variety of tools are available such as the Meyers-Briggs Type Indicator, DISC, Prevue, and others.
- Feedback from others. 360-degree-style tools that invite confidential feedback from a variety of different levels around us (peers, supervisors, and employees). Should be carefully supported with coaching.
- A mentor. Find an internal trusted colleague who you can turn to and ask for feedback.
- An external coach. A fresh set of eyes can push you with a bias for action and no burden to be politically correct or fear of upsetting you.
- Go to your mountain. Take time as a leader to disconnect and think. Use that time to "look in your mirror."

With self-reflection in hand (the data), we are best positioned to envision a future state (our success) and then work to craft a plan to get there. Without clarity about what we want as leaders, how to get there, the work we want to do, and why, highly effective leadership will be elusive. Self-reflection, clarity of purpose, and remembering that everyone is different might very well be the three-legged stool of leadership.

CHAPTER 18

Speaking Up and Out
Dan Ford

Blog Posts Written by Health Care Learners

To Be Obnoxious
Samantha Nino
June 16, 2014

"I'm a little obnoxious sometimes," claimed Dan Ford this afternoon while discussing his fervent advocacy of root-cause analyses in response to sentinel events. Earlier that morning, Mandy had confessed to being "that annoying nurse" who unabashedly telephones on-call residents when a concern arises. These champions of patient safety proudly own these deprecatory adjectives like *obnoxious* and *annoying* because they know that their actions are challenging the status quo for the betterment of patient care.

I hope that all of us, students and professionals alike, emerge from this week in Telluride a bit more enthusiastic about being obnoxious. To be *obnoxious* in this context is to put our patients' needs first in spite of a bruised ego. We *annoy* despite the fear of openly defying the medical culture's norms, and we *irritate* others because we have the courage to understand that it will take assertive individuals to lead the change that we need to see.

I worry that many of us, within the safe embrace of the Rocky

Mountains, can stand up and avow a career of advocacy for patient safety, transparency, and high reliability, but, as we return to our respective institutions, how many of us will succumb to the communities of fear that predominate our health care systems? How many of us just having completed the first year of medical school will have the fortitude to speak up to a resident during clinical rotations as a third-year medical student? We will all undoubtedly encounter scenarios that challenge our commitment to patient safety. Our egos should be attached to the idea of being the annoying physician, nurse, pharmacist, or medical student. We should be displeased with a presence that doesn't foster communication and teamwork within a care team. Standing out as an advocate and pioneer for patient safety should be the ideal; it should be our goal to be obnoxious.

Second Day at Telluride
Jameson Holloman
July 31, 2014

The past two days have been filled with some educational moments that have helped me gain personal insights into patient quality and safety. I'm hoping these experiences will serve as tools to make me a better physician in the future. We began yesterday with a film about Lewis Blackman, a fifteen-year-old patient who died as a result of a series of postsurgical medical errors. This film told the story from the perspective of Lewis's mother, Helen, who stayed with her son throughout Lewis's fumbled medical care in the hospital and tried desperately to save him from a system that was failing him.

One of the most powerful quotes in the movie was when Helen said (and I'm paraphrasing), "Had my son been anywhere but the hospital, he would be alive today."

The film put a face on the problem of medical error and humanized the issue. The need for a solution takes on a powerful impetus when you personally identify and empathize with the victims of medical error. It's one thing to look at statistics and analytics that describe the people affected by medical error, but when you can recognize the personal devastation that a lack of patient safety can cause a family, you feel much more empowered to act because you recognize the significance of change. This story served as a springboard to begin our exploration into patient quality and safety.

Our presentations, small-group discussions, and team activities examined different aspects of patient safety and quality. From these, two insights for me: in order to create a safer health care system, you need to first create a tiered system that anticipates and then corrects medical errors, and the system needs to be predicated upon open, honest, and effective communication between doctors, nurses, and patients.

Initial Reflections

Hospital clinicians and other staff often find themselves in a dilemma about their behavior and the behaviors of those around them. The negative behaviors range from isolated vignettes to dysfunctional environments in organizations with great to poor leadership. The quandary is: "Shall I speak up or not?" It takes courage and a strong value set when we are being tested to speak up and out. It also takes an understanding of the real value of communicating. The blog posts created by the TTE learners are perceptive, encouraging, thoughtful, refreshing, and thought-provoking. My sincere hope is that as they move along to their respective careers/ clinical practices, they will remember what they learned during their time with us.

The author of this chapter developed a deep passion for patient safety as a result of his wife's medical errors experienced in a hospital. The mother of three children (eleven, fourteen, and seventeen at the time) and age forty-seven, Diane was pursuing her second master's degree when she suffered a morphine-induced respiratory arrest following a scheduled hysterectomy, emergency colostomy, and other serious medical error harm. She has a permanent brain damage/short-term memory loss and a poor quality of life, and today, she resides in an independent-living facility.

The experience was traumatic for everyone. The family was not treated well when they asked logical and genuine questions about what happened. In 2003, the author was encouraged to start telling his wife's medical error story: The United States health care system has very serious medical errors problems. The national health care industry is lacking in credibility. Providers need to hear this story and other similar stories, over and over again. The author is now into his thirteenth year speaking about patient safety advocacy, passion, and journey of speaking up and out.

When I read *True North* (George, 2007), it was one of those a-ha moments for me. A key ingredient missing in "speaking up and out" is the lack of a reliance on one's true north, backbone, courage, and a supportive culture. Following this path when I am being seriously tested allows me to determine what is important to me and what is right—even if the decision is unpopular with others. True north is the internal compass that guides you successfully through life. It represents who you are as a human being at your deepest level. It is your orienting point—your fixed point in a spinning world—that helps you stay on track as a leader. Your true north is based on what is most important to you, your most cherished values, your passions, your motivations, and the sources of satisfaction in your life.

Just as a compass points toward a magnetic field, your true north pulls you toward the purpose of your leadership. When you follow your internal compass, your leadership will be authentic, and people will naturally want to associate with you. Although others may guide or influence you, your truth is derived from your life story—and only you can determine what it should be.

Discovering your true north takes a lifetime of commitment and learning. Each day, as you are tested in the world, you should look at yourself in the mirror and respect the person you see and the life you have chosen to lead. Some days will be better than others, but as long as you are true to who you are, you can cope with the most difficult circumstances that life presents (George & Sims, 2007).

Historical Background

Medicine and health care have long had a culture of not speaking up when bad things happen or when poor care is practiced. It's a culture of "cover up," according to Rosemary Gibson in her book, *Wall of Silence*. There are "club" cultures in which physicians make their way into and up the hierarchy because of their "fit" in the club and not because of their performance. Physicians who challenge the status quo and question suspect clinical outcomes can be viewed as disloyal. Unfortunately, some doctors choose to participate in the cover-up and do not report peers whose practice is substandard and a danger to the health of patients (Gibson, 2003). For many years, medical school and hospital medical and nursing staffs' cultures have stifled pushback or speaking up, a lack of mutual respect, and teamwork and have promoted individual decision-making, which sometimes takes the form of bullying.

Current State

Some progress is being made; more clinicians and staff are speaking up. However, there is still much reticence to speak up secondary to bullying and disrespectful behavior. A truly uncomfortable and/or stifling workplace can have a chilling effect on communication among staff. A survey completed by the Institute for Safe Medication Practices (ISMP) found that workplace bullying posed a critical problem for patient safety; rather than bring their questions about medication orders to a difficult doctor, almost half the health care personnel surveyed said they would rather keep silent. Furthermore, 7 percent of the respondents said that in the past year they had been involved in a medication error in which intimidation was at least partly responsible (ISMP, 2004).

In a recent blog post and personal communications with the author, John Self (2016), a respected nationally known health care executive search consultant, shares a current real-life example of lack of integrity in health care:

> A young interventional cardiologist, a star resident and cardiology fellow from a respected program, succumbed to an unbelievable offer of financial security from the competing hospital across town. She was assured that her new hospital was committed to quality, service, satisfaction, and doing the right thing for the patient. Three months into her tenure, she began to feel pressure to produce more revenue, more procedures.
>
> The pressure intensified as months passed. Though feeling trapped, she began to fudge here and there,

scheduling patients for procedures that she could justify in case anyone called her hand. She did not like doing this; she knew this was unnecessary care with a big unnecessary expense. But a few added procedures here and there helped ease the pressure for her employer. She surmised other cardiologists were doing the same thing. All the assurances she had received about quality of medicine and doing the right thing for patients appeared to go out the window once the ink was dry on her contract, which had onerous provisions regarding a noncompete. Everyone seemed to be turning a blind eye to her utilization numbers.

She was making great money, saving more than she thought possible for a growing family with expensive tastes. But the more money she made, the unhappier she became. In this story, the victim of integrity lapses questioned her new employers about bad behavior. She shares, "The common underlying theme in responses received in this: 'If we told you the truth, you might not have accepted the offer.'"

It is not only physicians who have difficulty speaking up. According to a study by Hendren (2011), nurses are involved in quality-improvement projects, committee work, hours devoted to preventing harm, checklists, protocols, and automated systems. However, poor communication is still the biggest patient safety danger of all. We have to wonder if we are focusing on the wrong metrics.

In an Association of periOperative Registered Nurses (AORN) report released in 2011, "The Silent Treatment," 85 percent of respondents said a safety tool had alerted them of a problem that might have been missed and potentially harmed a patient, yet 58 percent revealed they didn't say anything about it.

Among the study's key findings:

- More than four out of five nurses have concerns about dangerous shortcuts, incompetence, and disrespect demonstrated by their colleagues.
- More than half say shortcuts led to near misses or harm, and only 17 percent of those nurses shared their concerns with colleagues.
- More than a third say incompetence led to near misses or harm, and only 11 percent spoke to the colleague considered incompetent.
- More than half say disrespect prevented them from getting others to listen to them or respect their professional opinion, and only 16 percent confronted their disrespectful colleague.

Patients and families are speaking up, in part, because of The Joint Commission (TJC). The Patient Speak Up program at TJC launched it in March 2002. This was an encouraging sign for providers too. Activities of this nature can have a positive ripple effect as role models for clinicians speaking up and out (Joint Commission, 2002). The program urges patients to:

- Speak up if you have questions or concerns.
- Pay attention to the care you get.
- Educate yourself about your illness.
- Ask a trusted family member or friend to be your advocate (advisor or supporter).
- Know what medicines you take and why you take them.
- Use a health care organization that has been carefully checked out.
- Participate in all decisions about your treatment.

During a personal conversation with Mary Beth Kingston, executive vice president and chief nursing officer of Aurora Health Care in Milwaukee, the author discussed Speaking Up and how prepared the nurses are to advance this agenda.

Kingston acknowledged:

I'm not sure how well this is covered in nursing undergraduate education. The schools have definitely done a much better job of incorporating quality and safety concepts, but there is not a focus on Speaking Up. They need practice and reinforcement. In my opinion, nursing students rarely interact with physicians and others in their clinical rotations. They don't have the opportunity to learn these skills.

The reluctance to speak up has many facets, including hierarchies and lack of confidence. I'll often hear someone say, "But I wasn't certain ..." from the fear of retribution. It requires a concerted, clear message about the importance this plays in patient safety.

I think bullying does occur in the workplace from many sources, including lateral violence in nursing. Nurses will speak up, but as leaders, we need to ensure that we are supporting environments where speaking up is an expectation—and that everyone knows this and encourages it.

Nurses speak up more often now, but it is still a major issue. It is not a nursing issue alone. We have to ensure our culture supports this— meaning that everyone is not only prepared to speak up but to have someone speak up to them.

It is incumbent on leadership to address the issue of speaking up. Employees need to be empowered to speak up and not be treated disrespectfully or bullied. As CEO, Dr. John Toussaint introduced Toyota Production System principles to Thedacare, an integrated health system, in 2003. He is now CEO of an affiliated company, Center for Healthcare Value, and addresses this issue of speaking up through process improvement programs embedded into a hospital's culture and operations. The principle of "respect for people" means leadership respects caregivers enough to develop a system in which caregivers have control to change processes without management involvement. Management by process requires that leaders know exactly how care is delivered to patients and then lead frontline caregivers in improving those processes daily. The manager is not in control of everything that happens; he or she is a facilitator leading change. The system is designed to push decision-making to the smallest team at the frontline of the work.

Frontline teams have clear decision-making authority, but they can exercise it with the knowledge of what upper echelons of management are trying to accomplish overall. Executives must lead with humility, not power (Toussaint, 2015).

Lessons Learned

- To speak up and out for the benefit of the patient, patient safety, and quality, and for our colleagues.
- The need to challenge the status quo.
- Medical errors are considered the third-leading cause of death; concurrently, we have much difficulty speaking up. These dots must connect.
- Remember our own true north in every decisive moment.
- Learn from mistakes and near misses through speaking up.
- Patients and family members are role models for clinicians in this patient-focused/engagement culture.
- Leadership needs to spawn a culture in which everyone is free to speak up.
- Speaking up and out has to be a best practice in every organization, ingrained in every culture.
- Living and role-modeling your personal true north should be a personal best practice.
- Our egos should be attached to the idea of being the annoying physician, nurse, pharmacist, or medical student who does speak up and out.
- We should be displeased with a presence that doesn't foster communication and teamwork within a care team.
- Standing out as an advocate and pioneer for patient safety should be the ideal. It should be our goal to be obnoxious.
- Patients/family members are role models for speaking up and out and finding our true north.

The author was honored and proud to receive this communication from Alison Dolphin (TTE alumna) in November 2015:

> Hi Dan, I met you at TTE this summer and want to thank you for the wisdom you shared that has meant so much to me as a new graduate nurse: "Find your true north." In the midst of learning new techniques in patient care, I have made this my guiding light. When coerced to do something I wasn't comfortable with, I put on a stop, a pause, and remembered the goal of this work to help and not harm others, making patient safety paramount. Your words have given me strength, and I thank you for your attendance at that meeting and your support in meeting this common goal. Similarly, when I support other new nurses, I ask that they rise above getting clamored with guilt about the potential of error and mistake and aim to share and change practice so no one repeats their steps. This has meant so much to me that I want to share my deep gratitude.

CHAPTER 19

Gaming for Excellence
Anne Gunderson and Joseph Gunderson

Blog Posts Written by Health Care Learners

Gaming is a very important part of the TTE sessions. An attractive element of the gaming experience as a learning tool is that it provides opportunities for continued practice because negative consequences are not typically associated with failure. Rather, failure serves as an integral part of the learning experience. In traditional classroom settings, a learner who does not master a concept could be left with a gap in their knowledge foundation. This failure challenges later attempts to build to more complex concepts. In contrast, games inherently force the player to master a concept in order to advance. During our years at TTE, we have experimented with various games and exercises.

For example, we use a domino game that was created by members of our faculty. During the game, learners go through three rounds of dominoes. The learners are required to give and take instructions from their peers and have a short time to complete the task. Some players are only allowed to get directions without being able to ask questions. Others are allowed to ask questions. Some teams can ask questions but only have half the time to complete the task. Playing the roles as a nurse, physician, and observer allows the team to experience all the team feelings and reflections. According to the learners, this exercise helps them fully realize how limited and ineffective they can be in their daily interactions.

This is especially noted when it comes to communication. Learners note that it is very easy to get carried away and speak in our "default language." Unfortunately, their peers and the patients often get lost in the dark.

Health Care, Assemble!
Eva Luo
June 25, 2012

As our discussions on day 3 made very clear, health care professionals are not a united team. During a SBAR/I Pass The Baton communication exercise, it was blatantly obvious that, by the end of our first year of medical school, we have already been assimilated into our siloed profession. As medical students, we have absolutely no idea what nurses do or what their spheres of influence are in patient care.

As a fan of superhero movies, our deep confusion about how doctors and nurses work together to safely take care of patients reminded me of *The Avengers*. In the first half of the movie, a ragtag team of superheroes is brought together with the mission to save the world from alien domination. Each hero has his or her own superhuman talents (intelligence, strength, lightning-generating hammer, etc.). However, the villain's first attack quickly demonstrates that even superheroes are not effective alone and are at risk of causing unintentional harm. Thankfully, by the second half of the movie, the superheroes unite as a team, defeat the villain, and save Manhattan from complete destruction.

We had a few opportunities to challenge our communication and teamwork skills throughout the week. One of the most stressful exercises was a team challenge to balance a wooden plank on top of a cinderblock. Our task was to keep the plank balanced as six team members stepped on one by one, and then stepped off one by one. Sounds easy, right? Well, sitting under each end of the wooden plank was a "patient" (an egg). Could we load six team members onto the plank one member at a time and step off without "harming" our patients? The exercise demonstrated the importance of clear and precise communication and the need for situational awareness and agility to readjust plans as a team.

I was the first team member to step onto the plank, and the analogy of the egg as our patient was very real. My heart was racing throughout the entire exercise. My greatest concern was that my actions would throw off my team—and we would collectively hurt the patient. The faith in my team comforted me; I knew I had the power to stop the line, express my concerns, and get immediate feedback.

After nine minutes of tachycardia, my team triumphantly stepped off the plank without harming either "patient." (This is very hard to do). It was more proof of the power of effective teamwork. After the big win of saving the world from alien domination, the Avengers become a solidified team. They unite at the rallying cry of "Avengers, Assemble!" We need a similar rallying cry in health care. Let's start now!

Initial Reflections

As the learners continue to advance in this learning modality, so shall we. May the Force be with us!

Current State

Current medical, nursing, and other safety sciences belong to a generational cohort referred to as *millennials*. This cohort of learners is quite different than generations before. The gaming exercises that are utilized in the TTE curriculum provide some of the highest ratings from our learners. Due in part to their high degree of technological literacy, they are a radically different audience than the learners of fifteen years ago. Millennials enjoy learning about new technology through discovery, by experiencing and experimenting with it. They read less and are more comfortable in image-rich environments than with text. Their clear preference is for active, first-person, experiential learning and a level of interactivity that is absent in traditional lectures—but vibrantly present in new media technologies (Kron et al., 2010).

Advocates of game-based learning in higher education cite the ability of digital games to teach and reinforce skills important for future jobs, such as collaboration, problem-solving, and communication. In the Department of Education's 2010 National Education Technology Plan, there was a call for research in how "assessment technologies, such as simulations, collaborative environments, virtual worlds, games, and cognitive tutors, can be used to engage and motivate learners while assessing complex skills" (United States Department of Education, 2010, 15).

An attractive element of the gaming experience as a learning tool is that it provides opportunities for continued practice because negative consequences are not typically associated with failure. Instead, failure serves as an integral part of the learning experience (Gee, 2013; Groff, Howells, & Cranmer, 2010). This encourages learners to improve through repeated practice either by advancing within a game or replaying parts of a game.

Games meet the unique teaching and learning needs of learners when new concepts are introduced as a logical learning progression. Learning progressions are often described as the path learners take to learn a set of knowledge or skills (Forster & Masters, 1996). Learning progressions are frequently used in education.

Creating well-designed games is challenging but achievable. Games must challenge the learners and provide an obvious learning message. Games work best when coupled with effective pedagogy. As such, Steinkueler and Chmiel (2006) suggest that games will not replace teachers and classrooms, but they might replace some textbooks and laboratories. Games also meet the unique teaching and learning needs of learners when new concepts are introduced. Learners can take risks and quickly learn from their mistakes.

Learning does not end with the game. Games should be built with clear goals and provide immediate feedback to the learners. Debriefing is critical. Effective games provide feedback that is "(1) clear and unobtrusive, and (2) immediately responsive to the player's actions" (Ryan, Rigby, & Przybylski, 2006, 8). Effective debrief allows the leaners to connect what they learned in the gaming session and apply those concepts to other contexts. Educators can facilitate the transfer of skills by leading pregame and postgame discussions that connect the game with other things students are learning in class (Ash, 2011).

Learners can be encouraged to share different ways of approaching a problem. This allows learners to change their strategies in order to improve their performance. The idea of immediate feedback is also prominent in good formative assessment processes. Ideally, learners will improve when given constructive feedback.

During TTE, we use multiple games to reinforce the content learned via traditional education. The games are designed to engage the participants in interactive learning. The games are often the highlight of the week for many learners. They might forget the presentations they observed, but they never forget the games. Ask a learner who was on the teeter-totter during one of our gaming simulations and broke the eggs, which represent the "patients," and they will tell you that they were devastated. They will admit to begging for a second chance to right their wrong and save their "patients." The emotions and internal growth is palpable and observable.

We also use a domino game to address issues in communication across providers. While we don't bring the patients into play, we do get to a raw and real issue in health care: the terrible communication between providers. Even better, in five minutes of using the snowflake paper game, we can prove that, even when they all hear the same directions at the same time, they will still end up with different outcomes based on their first fold.

The magic of gaming comes alive at TTE and burns brightly in the lives and practices of our alumni. These low-budget games are quick, easy to administer, and deliver impactful and lasting lessons. You can find a variety of these games easily online or contact us at TTESupport@ medstar.net for the safety and quality gaming we use.

Lessons Learned

- Creating well-designed games is challenging but achievable.
- In traditional learning settings, a learner who does not master a concept often gets left behind in their knowledge foundation.
- Effective debrief allows the learners to connect what they learned in the gaming session and apply those concepts to other contexts.
- Using gaming to master difficult concepts can assist learners in their advancement.

- Learners need to assimilate with all health care learners—not just with those in their own silos.
- Working in teams is critical to success; gaming supports this learning modality.
- The gaming part of TTE provides learning that can't be gained by reading a book or watching a PowerPoint lecture.
- Nobody who breaks an egg (a patient) ever forgets it.

CHAPTER 20

The Eternal Flame
Paul Levy

At Arlington National Cemetery

The TTE participants in the Maryland session have an opportunity to reflect on their experiences when the group visits Arlington National Cemetery. The memorial is dedicated to those who have served in the armed forces of the United States after 1864 and is a physically impressive space—624 vast acres, silent, and otherwise like a pleasant park with rolling hills and ancient trees—but dotted with the shining white gravestones of four hundred thousand people.

On a hill overlooking the cemetery, just above John F. Kennedy's burial site, our participants are first provided with a history of the cemetery. Regardless of national origin, they are drawn in silent admiration and sadness for what they see in front of them. They understand that the best of a nation's youth from across many decades are in the cemetery because of their commitment to the country's standards of freedom and equality—a commitment that led to their service and ultimate death. Those young men and women knew what they were facing, and they knew that they might never see their loved ones again.

How different for patients and families who experience the ravages of the health care system. In one year, roughly the same number of people contained in Arlington Cemetery die unnecessarily while in the care of America's hospitals. Comparable numbers apply to other countries.

Simone Erchov (TTE alumna) reflects on this in her blog post "Impact Through Metaphors":

> An incredibly strong, sobering metaphor was made today at Arlington National Cemetery. The notion that four hundred thousand lives lost over the course of several wars in the pursuit to protect the freedom of citizens to pursue meaningful goals matches the loss of those citizen lives every year due to medical error is disheartening and baffling. It hits home because just as each member in this cemetery belongs to a family … leaves behind loved ones … each patient who loses his or her life does too.

Rosemary Gibson offers the learners an excerpt from *The Wall of Silence*. It is a letter written by a mother, Ilene Corina, to a doctor after the preventable death of her child:

> I really don't know why I am writing this. For some reason, I feel the need to. I don't hate you. You were a nice man when I met you, and I'm sure you're a nice man now.
>
> My life was torn apart when my son died and can never fully be put back together again. Michael was my firstborn, my only child, he was my whole life. Through carelessness, he was taken away forever. I've spent many nights and awakened many mornings wondering if I should go to be with my son. I wonder if I'll ever stop feeling this way.
>
> I wonder often if you were leaving for vacation or had a dinner engagement that you didn't want to miss, if this is why Michael wasn't checked further.
>
> Were you too busy? I'm sure you're a good doctor. But the saying is a baker eats his mistakes and doctors bury theirs.
>
> I know your life isn't going to be disrupted by my tragedy, and I want your children to always feel proud

that their daddy is a good doctor. But I only pray that every mother who brings her child to you is a reminder that that child is a piece of her heart and you think of my son, Michael, and my heart.

Bryan LaBore preferred to find hope from his visit to Arlington, even while citing one of the benchmark studies about the failures of the American health care system in his blog "The Eternal Flame":

"To Err Is Human" is not a parade pamphlet but an acknowledgement of defeat. We shouldn't need four hundred thousand tombstones surrounding us to realize that. I think that is the thing about this conference that gives me the most hope. We are thirty young medical professionals who are ready to declare war on an unsafe system, but we don't quite know how to fight. We have a burning desire to change medicine for a better. It's an eternal flame.

Erchov suggests that there is hope if doctors and other clinicians come to understand that their profession demands a discipline comparable to those who engage in the changing of the guard:

Medical professionals dedicate their lives in a noble pursuit just as the sentinels do, so perhaps there are more lessons to glean from these teams. The lessons are not only in the precision of their work but the driving force behind it—the reminder of the cause. They are not allowed to forget; their consistent pre- and post-shift time-out offers and demands that much. The mind-set toward such a task should therefore be, "This is a gift that gives us a chance to regroup and get ourselves on track" rather than "This is a nuisance protocol through which we must rush due to time constraints." Understanding the purpose of these checks, the value of teamwork, and a common language, the power of the interdisciplinary

224

makeup—these are the lessons of which such a ritual can remind us. Each move we make is out of respect for the life of the patient just as the sentinel's is for the loss of the unknown who dedicated their lives to protecting the freedom to become the professionals we are today and will be in the future.

The paradox we face in the American health care system is this: How can it be that the cadre of doctors—well-intentioned, highly trained, and intelligent doctors—collectively represent one of the highest-ranked public health hazards in the country when they work together in our hospitals? Erchov hints at part of the answer. There is a predisposition— whether because of training or self-selection or more—for doctors to believe that they should apply their creativity, judgment, and intuition when taking care of patients. Of course, clinical variation is necessary; bodies and medical conditions come in lots of shapes and sizes. It would be absurd for us to insist that doctors treat every person in exactly the same way. We must rely on the training, experience, judgment, and creativity of doctors to recognize when variation from a protocol is required.

Does that suggest that there is no place for reducing variation in practice? Hardly. The mere existence of wide variation in some clinical processes is, in itself, a major cause of harm in the world's hospitals. You can scarcely be practicing evidence-based medicine if the degree of clinical variation exceeds that which is thoughtfully required. Also, it is difficult to engage in systemic process improvement if there is an insufficient baseline against which to compare the results from experimental steps in improving any given process.

In the drive to reduce preventable harm and improve quality of care, where and how should we employ protocols? How do we introduce such concepts in a manner that will be endorsed and accepted by the vast majority of clinicians? It is essential that we do so in a manner that reflects the value judgments and predispositions of doctors, teaching them how to employ the scientific method in bringing about clinical process improvement. Doctors are, at heart, scientists with respect for the power of data and analysis.

One approach summarized by Brent James at Intermountain Health is a process known as "shared baselines," which is used in Intermountain Health System. It has several components:

1. Select a high-priority clinical process.
2. Create evidence-based best practice guidelines.
3. Build the guidelines into the flow of clinical work.
4. Use the guidelines as a shared baseline—with doctors free to vary them based on individual patient needs.
5. Meanwhile, learn from and (over time) eliminate variation arising from the professionals, while retain variation arising from patients.

It is important to understand that this approach *demands* that doctors modify shared protocols on the basis of patient needs. The aim is not to step between doctors and their patients. However, this is very different from the free form of patient care that exists generally in medicine. Brent, noted, "We pay for our personal autonomy with the lives of our patients. This is indefensible." In short, the approach used at Intermountain values variation based on the patient and not the physician.

It would be too simple to extrapolate from this concept and head to a strong reliance on checklists to embody the protocols, but that would be too simplistic because the environment within which checklists are carried out has a strong impact on their effectiveness.

To be truly effective, a checklist must be a tool that is wisely used among a group of people. Flight Captain Chesley "Sully" Sullenberger reminds us: "A checklist alone is not sufficient. What makes it effective are the attitude, behavior, and teamwork that go along with the use of it."

In that regard, Sully has noted the importance of crew resource management (CRM) techniques. He has defined CRM as "a compact, with defined goals and responsibilities" among team members. He notes, "These are not soft skills. They are human skills. They have the potential to save more lives than new medical technologies."

And so, at Arlington, we come full circle: The emotional impact of so much death is palpable among the TTE learners. Their commitment to change is strong. The tools they have been given can be effective. Now their task is to go back to their home institutions and, day by day, make incremental improvements in communication, mindfulness, clinical process improvement, resilience, and reliability. They have mentors and allies they can call upon, whether among the faculty and alumni of TTE or among their colleagues in the field.

CHAPTER 21

Education Solutions
Anne Gunderson and Wendy Madigosky

Blog Posts Written by Health Care Learners

A Burning Platform
Eva Luo
June 19, 2012

It is of utmost importance to put the patient rather than one's self at center stage. The task that resonated most with me was a reminder to keep the patient at the center.

- Lead reform in putting the patient first.
- Be a motivator, mentor, and facilitator.
- Be a communicator.
- Be a team leader and a team player.
- Demonstrate high-level clinical skills and research.
- Manage finances.
- Think critically.
- Monitor one's own performance.
- Behave in an honest, open, and ethical manner.
- Display integrity.
- See the big picture.
- Be able to learn from experience.

Twenty Lessons Learned in TTE
Heidi Charvet
June 17, 2013

I just returned from an amazing week in TTE where I learned a lot and was inspired and reinvigorated by the group of colleagues and faculty I met. One of the most surprising things about the week was that, despite our varied backgrounds and geography, we all came together with a common set of interests and experiences (sometimes bad ones) determined to make changes moving forward in our practice:

1. Start every meeting with a story (it's all about the patient!).
2. The way we treat nurses when they bring a concern to us that ends up being wrong is often more important than when they bring a concern that ends up being right. Respect and appreciation are essential—or else you may not hear about the next concern that ends up harming a patient.
3. The art of negotiation is about "letting them have your way."
4. Culture eats strategy for breakfast.
5. Patients who end up in the "wrong ward" are often the most vulnerable.
6. Your greatest enemy is complacency.
7. Informed consent is a process—not an event. An even better term for it is *shared decision-making*.
8. Doctor language can be disrespectful to patients (whoa!).
9. Problem-solving is one of the main barriers to listening among doctors. We need to be cognizant of this when we find ourselves tuning out.
10. Once you have made your point, shut up and listen.
11. Seek first to understand and then be understood.
12. Holding "patient safety event reviews" focused on resident education rather than root-cause analysis may be a better way to achieve buy-in.
13. If you want to report something, but the other person involved is resisting, write it together so there will be no misunderstandings or division of blame.

14. Reporting a patient safety concern is about the *incident* (and the harm or potential to harm) not the individuals involved.
15. There is a role for situational leadership in health care. You lead differently on a sinking ship than when making changes about parking at the hospital.
16. Leaders do not make excuses—they build solutions.
17. Some is not a number; soon is not a time.
18. As physicians, sometimes we take steps to treat ourselves and not the patients.
19. Build trust by saying what you are going to do and then doing it.
20. Knowledge is power. Your job is to transfer this power to the patient.

Initial Reflections

Just reading the blog posts for this chapter makes it obvious that TTE has been wildly successful; it's possible that we don't even need faculty to address this last chapter. As we reread the blog posts, we are humbled by the amazing teaching and learning that has gone on since we started inviting learners to TTE. We are blessed with dedicated and very talented faculty who make the TTE sessions possible. This chapter is focused on what needs to happen in the United States to ensure all health science learners are trained at the same level our current TTE alumni are trained. The learners have said it best.

Historical Background

Patient safety and quality management are essential disciplines within health care education (Gunderson, 2008). Safety science aims to achieve a trustworthy system of health care delivery by minimizing the incidence and impact of adverse events and maximizing recovery when they do occur. The methods of quality management come largely from disciplines outside medicine, particularly from cognitive psychology, human factors engineering, and organizational management science.

Dr. Jordan Cohen, former president of the Association of American Medical Colleges, once stated there needed to be a "collaborative effort to ensure that the next generation of physicians is adequately prepared to recognize the sources of error in medical practice, to acknowledge their own vulnerability to error, and to engage fully in the process of continuous quality improvement" (1999). Health care professionals with a broad understanding of these subjects will thus be better equipped to lead successful quality-improvement initiatives. Unfortunately, although there has been some advancement in safety and quality curricula, a formal systematic patient safety and quality curriculum remains largely absent within health care education. For this reason, several innovative health care educators are currently creating ways to protect and empower learners—and keep patients safe—through novel patient safety education opportunities.

Experts in patient safety and quality management determined that achieving and sustaining broad-based access to high-quality, affordable care requires understanding and elimination of major structural barriers that impede the development of a high-performing American health care system. These barriers can either be constructed or dismantled by a system's culture, which according to the Institute of Medicine (IOM) is central to promoting learning at every level. The need for creating a new culture of care is common to all types of health care organizations; all need to cultivate the ability to continuously learn and improve in order to prevent medical error from remaining a leading cause of death.

Current State

TTE, created in 2009, provides learning opportunities for novice learners in patient safety, leadership, and quality improvement. The vision is to design, create, and provide exceptional training for medical and nursing learners committed to becoming future leaders in helping reduce medical error and preventable harm. The ultimate goal of the TTE curriculum is to alter the current culture of error and create continuously evolving organizations that are capable of generating and utilizing knowledge from every patient interaction to yield greater performance, predictability, and reliability. In 2016, approximately 240 learners attended TTE globally.

The program curriculum has four objectives: (1) highlight the importance of open, honest, and effective communication between caregivers and patients for improving patient safety; (2) recognize and apply basic communication skills toward improving effective communication among members of the health care team; (3) utilize tools and strategies to lead change specific to reducing patient harm; and (4) implement, lead, and successfully complete a safety or quality improvement project at the learners' home institution in the next twelve months. The hope is that the interprofessional education lessons learned from TTE can be expanded to provide necessary training to all medical and nursing students throughout the United States and ideally internationally.

Overall, the learners desire fundamental understanding, advanced discussion, and critical evaluation of patient safety and quality improvement. Alumni networking, story sharing, and deliberate practice of project-development strategies better educate learners and allow them to act based on the lessons learned during TTE. Learners are currently provided with pre-reading materials designed to help them begin thinking about patient safety and quality-improvement topics prior to arriving. These materials serve to pique learners' interests rather than provide education in specific, deliberate patient safety and quality-improvement topics.

The interprofessional aspect of TTE emphasizes the reality of the hospital setting for the health students. Collaborative and team-based efforts among health professionals correlate highly with safe and effective health care. Effective interprofessional teamwork is known to reduce errors caused by miscommunication and poor patient care handover. Early education environments that emphasize interprofessional learning will better prepare learners for the work environment they are entering or of which they are already a part. Continuous improvement is also taught during the TTE sessions. Health care practitioners require systematic problem-solving skills and the application of systems-engineering techniques. It is essential to have operational models that encourage and reward sustained quality improvement, transparency on cost and outcomes, and strong leadership with a vision devoted to improving health care processes. Health care professionals have a tremendous need to develop expertise in the fields of health care quality and patient safety.

TTE provides learners with interprofessional education in patient safety and quality improvement, but there is still need for further curriculum refinement. Successful training programs require frequent evaluation. Based on a recent qualitative and quantitative survey completed by TTE learners, we have identified improvements that will improve the curriculum in three key areas: (1) more small group discussion built into the patient safety curriculum, (2) greater balance in the quality-improvement curriculum between identification of problem areas and strategies for implementation of solutions, and (3) increased networking opportunities and story sharing among participants. Each of these three areas requires a reduction in lecture-based learning with increased time for learners to interact in a smaller group-based setting.

The learners believe that making these changes will ensure time for a basic and uniform understanding of the topics among the learners and provide more time for productive conversations. Moving forward, the learners suggested that the faculty should provide participants with specific pre-conference objectives and learning material in the form of articles and online lectures. This would provide guidance and a means for self-directed learning among the learners, hopefully leading to more enriching discussions during the training. Finally, the learners strongly believe a model must be developed for large-scale education in these subjects. TTE is currently able to train 240 students globally each year, but it will take the work of our entire health care system to make significant progress in reducing the great harm caused by medical error.

Lessons Learned

The development and implementation of patient safety and quality-improvement learning opportunities is important to the health care community in its continued efforts to break the cycle of medical error. The goal of every health science institution should be to facilitate dynamic educational opportunities in a learner-centered environment to enhance learners' knowledge, attitudes, and skill development. The curriculum should emphasize training in the major fields of patient safety, quality improvement, communication, collaboration, error science, and process improvement.

As future leaders in the health care arena, health science learners should be provided with the opportunity to learn how to participate and take leadership roles in a broader interprofessional community that includes basic and social scientists, patients, clinicians, allied health professionals, health care leaders, policy makers, licensing agencies, and accrediting bodies. Most important, the education should be designed to facilitate a culture of safety and increase the ability of health sciences learners to become leaders in the health care system. High-risk, safety-oriented industries such as aviation, nuclear energy, and behavioral science can provide successful safety and quality curriculum-innovation strategies.

Without competent patient safety and quality-improvement education, medical providers will continue to fall prey to the numerous safety and quality pitfalls in the health care environment. To change the current culture to a culture of optimal quality and safety, it is vital that learners understand and appreciate the direct link between scientific advances and improving health care outcomes before moving on to clinical practice. A variety of teaching strategies can result in better learning, retention, and positive application of learned information by each learner. Successful facilitation of adult learning requires well-thought-out, evidence-based content presented in a format that is appealing to a diverse group of learners. This is not an easy task, but it must be done.

As noted by Yukl, "A vision is seldom created in a single moment of revelation, but instead it takes shape during a lengthy process of exploration, discussion, and refinement of ideas" (2002).

ACKNOWLEDGMENTS

Authors

Jan Boller PhD, RN
Richard Corder, MHA, FACHE
Dan Ford, MBA, LFACHE
Lisa Freeman, BA
Robert M. Galbraith, MD, FACP
Rosemary Gibson, MSc
Stacey M. Gonzalez, BA
Tracy Granzyk, MS, MFA
Anne J. Gunderson, EdD, MS,GNP
Jessica A. Gunderson, MS
Joseph D. Gunderson, BFA
Helen Haskell, MA
Carole Hemmelgarn, MS
Roger Leonard, MD, MMM, FACC, CPE
Paul F. Levy
Sherri T. Loeb, RN, BSN
Joan Lowery, MEd
Wendy S. Madigosky, MD, MSPH
David B. Mayer, MD
Cathy Mayer, RN
Timothy B. McDonald, MD, JD
Farzana S. Mohamed
Armando Nahum
John J. Nance, JD
Kim Oates, MD, DSc, MHP, FRACP

Katherine A. Pischke-Winn, MS, MBA, RN
Gwen Sherwood, PhD, RN, FAAN, ANEF
Jacqueline T. Stark, BS
Dan Walters

Learners

Amishi Bajaj
Maureen Baker
Christine Beeson
Mona Beier
Derek Blok
Erin Bredenberg
Carey Candrian
Nicholas Clark
Garrett Coyan
Cassidy Dahn
Allison Dolphin
Mihai Dumitrescu
Simone Erchov
Elissa Falconer
Caitlin Farrell
Christian Gausvik
Alana Gilman
Ann Harrington
Jameson Holloman
Eghosa Isa
Cassie Jia
Bryan Labore
Eva Luo
Nicole Martin
Suresh Mohan
Julie Morrison
Heidi Nicewarner
Samantha Nino
Michael Parker

Colleen Parrish
Rupa Prasad
Padma Ravi
Aubrey Samost
Laura Schapiro
Daniella Schocken
Junzi Shi
Evan Skinner
Giovanna Sobrinho
Kristin Streiler
Lakshman Swamy
Christine Thomas
Nicole Werner
Christopher Worsham

OUR LANGUAGE

ACGME. The American College of Graduate Medical Education.

ADA. American Disabilities Act.

adverse event. Any untoward medical occurrence in a patient or clinical investigation subject administered a pharmaceutical product and which does not necessarily have a causal relationship with this treatment.

AELPS. Academy for Emerging Leaders in Patient Safety.

AHRQ. Agency for Healthcare Research and Quality.

AORN. Association of periOperative Registered Nurses.

BMJ. British Medical Journal.

care for the caregiver. Providing support for a caregiver after a harm event occurs.

Carlos Castaneda. Carlos Castaneda was an American author with a PhD in anthropology who wrote a series of books that describe his training in shamanism.

CBC. complete blood count.

CLABSI. Central line-associated bloodstream infection.

clerkships. Clinical clerkships encompass a period of medical education in which students—medical, nursing, dental, or otherwise—practice medicine under the supervision of a health practitioner.

CMS. Centers for Medicare and Medicaid Services.

COPD. Chronic obstructive pulmonary disease.

CSF. Cerebrospinal fluid.

C-suite. A widely used slang term used to collectively refer to a corporation's most important senior executives, e.g. chief executive officer.

CT. Computed tomography. A CT scan makes use of computer-processed combinations of many X-ray images taken from different angles to produce cross-sectional images of specific areas of a scanned object, allowing the user to see inside the object without cutting.

culture. The integrated pattern of human knowledge, belief, and behavior that depends upon the capacity for learning and transmitting knowledge to succeeding generations; the set of shared attitudes, values, goals, and practices that characterizes an institution or organization <a corporate culture focused on the bottom line>; the set of values, conventions, or social practices associated with a particular field, activity, or societal characteristic (Merriam-Webster.com, n. d.).

da Vinci Robot. The da Vinci Surgical System enables surgeons to perform operations through a few small incisions and has several key features.

disclosure. Health privacy; the release or divulgence of information by an entity to persons or organizations outside of that entity.

Fio2. Fraction of inspired oxygen (FiO2) is the fraction or percentage of oxygen in the space being measured. Medical patients experiencing difficulty breathing are provided with oxygen-enriched air, which means a higher-than-atmospheric FiO2. Natural air includes 20.9 percent oxygen, which is equivalent to FiO2 of 0.209. Oxygen-enriched air has a higher FiO2 than 0.21, up to 1.00, which means 100 percent oxygen.

GME. Graduate medical education.

go team. A group of investigators who can be dispatched immediately to investigate accidents, attacks, and the like.

good catch. An event that was discovered or caught before it reached the patient, that if it had not may have resulted in harm.

HAI. Hospital-acquired infection.

HCAHPS. Hospital consumer assessment of healthcare providers and systems.

HRO. High-reliability organization.

ICU. Intensive care unit.

IHI. Institute for Healthcare Improvement.

informed consent. The process by which the treating health care provider discloses appropriate information to a competent patient so that the patient may make a voluntary choice to accept or refuse treatment.

IOM. Institute of Medicine.

IT. Information technology.

IVC. INFERIOR vena cava.

JHACO. Joint Commission on the Accreditation of Healthcare Organizations.

just culture. A term that refers to a culture that is both fair to staff who make errors and effective in reducing safety risks. In a just culture, all staff know that safety is valued in the organization, and they continually look for risks that pose a threat to safety.

LEAP. Love, energy, audacity, proof.

MBSR. Mindfulness-based stress reduction.

MRI. Magnetic resonance imaging.

NCLEX. National Council Licensure Examination is the exam that nurses take to become licensed in the United States.

NEJM. New England Journal of Medicine.

NHS. National Health Service.

NPSF. National Patient Safety Foundation.

opaqueness. Difficult to understand; not transparent.

PFACQS. Patient Family Advisory Council on Quality and Safety.

plaintiff's attorney. A plaintiff's attorney is a lawyer who represents individuals who have been harmed physically or financially. They fight for the rights of the "little guy" against the powerful. Plaintiffs' attorneys typically take on corporations, insurance companies, hospitals, business interests, and even governmental organizations.

QSEN. Quality and Safety Education in Nursing.

Rapid-response teams. Failures in planning and communication, and failure to recognize when a patient's condition is deteriorating, can lead to failure to rescue and become a key contributor to in-hospital mortality. If identified in a timely fashion, unnecessary deaths can often be prevented. The rapid-response team—known by some as the medical-emergency team—is a team of clinicians who bring critical care expertise to the bedside. Simply put, the purpose of the rapid-response team is to bring critical care expertise to the patient's bedside (or wherever it's needed).

RCA. Root-cause analysis.

second victim. Care for frontline workers who are involved in or exposed to an adverse event.

sepsis. A potentially life-threatening complication of an infection. Sepsis occurs when chemicals released into the bloodstream to fight the infection trigger inflammatory responses throughout the body and can cause organ failure.

shared decision-making. A process in which clinicians and patients work together to make decisions and select tests, treatments, and care plans based on clinical evidence that balances risks and expected outcomes with patient preferences and values.

SNF. Skilled nursing facility.

TTE. The Telluride Experience.

VAP. Ventilator-associated pneumonia.

WBC. White blood count.

workarounds. A method for overcoming a problem or limitation in a program or system.

GLOSSARY OF TITLES

ANEF. Academy of Nursing Education Fellow
BA. Bachelor of Arts
BSN. Bachelor of Science in Nursing
CMA. Certified Medical Assistant
CNA. Certified Nursing Assistant
CNM. Certified Nurse Midwife
CPE. Clinical Pastoral Education
DC. Doctor of Chiropractic
DDS. Doctor of Dental Surgery
DMD. Doctor of Dental Medicine
DNP. Doctor of Nursing Practice
DO. Doctor of Osteopathy
DPT. Doctor of Physical Therapy
DSc. Doctor of Science
EdD. Doctor of Education
EMTB. Emergency Medical Technician, Basic
EMT-P. Emergency Medical Technician, Paramedic
FAAN. Fellow of the American Academy of Nursing
FACC. Fellow of the American College of Cardiology
FACHE. Fellow of the American College of Healthcare Executives
FACP. Fellow of the American College of Physicians
GNP. Gerontological Nurse Practitioner
JD. Doctor of Jurisprudence (Doctor of Laws)
LPN. Licensed Practical Nurse
MA. Master of Arts
MBA. Master of Business Administration
MD or DM. Doctor of Medicine

MEd. Master of Education

MFA. Master of Fine Arts

MHA. Masters in Health Administration

MHP. Master of Health Planning

MMM. Master of Medical Management

MPH. Master of Public Health

MS. Master of Science

MSPH. Master of Science in Public Health

ND. Doctor of Naturopathy

NP. Nurse Practitioner

PA-C. Physician Assistant Certified

PD or PharmD. Doctor of Pharmacy

PhD. Doctor of Philosophy

PT. Physical Therapist

RN. Registered Nurse

REFERENCES

Agency for Healthcare Research and Quality. March 2014. Chapter 5: Overall Results. Retrieved from http://www.ahrq.gov/professionals/quality-patient-safety/patientsafetyculture/hospital/2014/hosp14ch5.html.

Agency for Healthcare Research and Quality. n. d. "Communication and optimal resolution (CANDOR) toolkit." Retrieved from http://www.ahrq.gov/professionals/quality-patient-safety/patient-safety resources/resources/candor/introduction.html.

Agency for Healthcare Research and Quality. n. d. "How patient and family engagement benefits your hospital." Retrieved from http://www.ahrq.gov.

American Academy of Pediatrics. 2012. "Patient-and-family-centered care and the pediatrician's role." *Pediatrics* 129 (2): 394–404. doi:10/1542/peds.2011–308.

Ash, K. 2011. "Digital gaming goes academic." *Education Week* 30 (25): 24–28.

Banja, J. D. 2005. *Medical Error and Medical Narcissism.* Sudbury, MA: Jones and Bartlett Publishers, Inc.

Bass, B. M. & Riggio, R. E. 2006. *Transformational Leadership* (2nd ed.). Mahwah, NJ: Erlbuam.

Becher, E. C., & Chassin, M. R. 2001. "Improving the quality of health care: Who will lead?" *Health Affairs,* 20 (5), 164–179.

Bellot, J. 2011. "Defining and assessing organizational culture." *Nursing Forum,* 46 (1), 29–37.

Berk, J. 2013. "The 5 secrets of storytelling for social change." *Forbes.* Retrieved from http://www.forbes.com/sites/

skollworldforum/2013/08/01/the-5-secrets-of-storytellin
g-for-social-change/.

Berwick, D. December 2015. Speech: "Turtles," IHI national forum
plenary speaker 2015, and personal communications with D.
Ford.

Bishop, S. R., Lau, M., Shapiro, S., Carlson, L., Anderson, N. D., Carmody,
J., Segal, Z. V., Abbey, S.,Speca, M., Velting, D., & Devins, G.
2004. "Mindfulness: A proposed operation definition." *Clinical
Psychology: Science and Practice*, 11 (3), 230–241.

Bismark, M., Spittal, M., Gurrin, L., Ward, M., & Studdert, D. April 10,
2013. "Identification of doctors at risk of recurrent complaints: A
national study of health care complaints in Australia." *BMJ Quality
and Safety*. Retrieved from http://qualitysafety.bmj.com/content/
early/2013/02/22/bmjqs-2012-001691.full.pdf+html.

Boothman, R. C., A. C. Blackwell, D. A. Campbell Jr, E. Commiskey, &
S. Anderson. 2009. "A better approach to medical malpractice
claims? The University of Michigan Experience." *Journal of Health
& Life Sciences Law* 2 (2): 125–59.

Brennan, T., Leape, L., Laird, N., Hebert, L., Localio, R., Lawthers, A.,
Newhouse, J., Weiller. P., & Hiatt, H. Feb. 7, 1991. "Incidence of
adverse events and negligence in hospitalized patients—Results
of the Harvard Medical Practice study." *The New England Journal
of Medicine*.

Briskin, A., Erickson, S., Ott, J., & Callanan, T. 2009. *The Power of Collective
Wisdom and the Trap of Collective Folly*. San Francisco, CA: Berrett-
Koehler Publishers, Inc.

Brown, T. June 2008. Design thinking. *Harvard Business Review*, 1–9.

Brown, T. May 7, 2011. "Physician, heal thyself." *The New York Times*.
Retrieved from: http://www.nytimes.com/2011/05/08/
opinion/08Brown.html?_r=0.

Brown, G., Bifulco, A., Harris, T., & Bridge, L. 1986. "Life stress, chronic
subclinical symptoms and vulnerability to clinical depression."
Journal of Affective Disorder 11 (1). Retrieved from http://www.jad-
journal.com/article/0165-0327(86)90054-6/pdf.

Brown, K. W., Ryan, R. M., & Creswell, J. D. 2007. "Mindfulness: Theoretical foundations and evidence for its salutary effects." *Psychological Inquiry* 18 (4): 211–237.

Burdine, J., Felix, M., Abel, A., & Musselman, Y. October 2000. "The SF-12 as a population health measure: an exploratory examination of potential for application." *Health Services Research* 35 (4). Retrieved from https://www-ncbi-nlm-nih-gov.proxy.library.georgetown.edu/pmc/articles/PMC1089158/.

Carnett, W. G. 2002. "Clinical practice guidelines: A tool to improve care." *Journal of Nursing Care Quality* 16 (3): 10.

Cha, A. E. May 3, 2016. "Researchers: Medical errors now third leading cause of death in United States." *The Washington Post*. Retrieved from https://www.washingtonpost.com/news/to-your-health/wp/2016/05/03/researchers-medical-errors-now-third-leading-cause-of-death-in-united-states/.

Charon, R. & Hermann, N. 2012. "Commentary: A sense of story, or why teach reflective writing?" *Academic Medicine, 87,* 5–7.

Cleveland Clinic. Feb. 2014 "Empathy: The human connection to patient care" [YouTube video]. Retrieved from https://www.youtube.com/watch?v=cDDWvj_q-o8.

Cohen, J. Dec. 1999. *Letter to medical school deans.* Washington, DC: Association of American Medical Colleges.

Cooke, M., Irby, D. M., Sullivan, W., & Ludmerer, K. M. 2006. "American medical education 100 years after the Flexner report." *New England Journal of Medicine, 355,* 1339–1344.

Conway, J., Johnson, B., Edgman-Levitan, S., Schlucter, J., Ford, D., Sodomka, P., & Simmons, L. 2006. *Partnering with patients and families to design a patient- and family-centered health care system: A road map for the future.* Bethesda, MD: Institute for Family-Centered Care.

Cox, LM & Logio, L. S. 2011." Patient safety stories: A project utilizing narratives in resident training." *Academic Medicine* 86 (11): 1473–1478.

Cron, L. 2012. *Wired for Story: The Writer's Guide to Using Brain Science to Hook Readers from the Very First Sentence.* Berkeley: Ten Speed Press.

Dall'ra, C., Griffiths, P., Ball, J., Simon, M., & Aiken, L. Aug. 23, 2015. "Association of 12 h shifts and nurses' job satisfaction, burnout and intention to leave: Findings from a cross-sectional study of 12 European countries." *BMJ Open, 5.* Retrieved from http://bmjopen.bmj.com/content/5/9/e008331.full.pdf+html.

Dhaliwal, G. March 16, 2016. "If only health care would focus on this one thing." *The Wall Street Journal.* Retrieved from http://blogs.wsj.com/experts/2016/02/16/if-only-health care-would-focus-on-this-one-thing/.

Djikic, M., Oatley, K., Zoeterman, S. & Peterson, J. B. 2009. "On being moved by art: How reading fiction transforms the self." *Creativity Research Journal* 21 (1): 24–29.

Dobkin, P. L., & Hutchinson, T. A. 2013. "Teaching mindfulness in medical school: Where are we now and where are we going?" *Medical Education* 47, 768–779.

Dolphin, A. Nov. 11, 2015 & Feb. 8, 2016. Personal Communications with D. Ford.

Doss, H. May 23, 2014. "Design thinking in health care: One step at a time." *Forbes.* Retrieved from http://www.forbes.com/sites/henrydoss/2014/05/23/design-thinking-in-health care-one-step-at-a-time/#283a7295628d.

Dossey, B. M. 2010. *Florence Nightingale: Mystic, visionary, healer.* Philadelphia, PA: F. A. Davis Company.

Dreyfus, H. L. & Dreyfus, S. 2008. "Beyond expertise: Some preliminary thoughts on mastery." In K. Nielsen (Ed). *A Qualitative Stance: Essays in Honor of Steiner Kvale,* 113–124. Aarhau University Press.

Dyerbye, L., Thomas, M., Eacker, A., Harper, W., & Massie Jr., S. (Oct. 22, 2007). "Race, ethnicity, and medical student well-being in the United States." *Archives of Internal Medicine,* 167 (19), 2103–2109. doi:10.1001/archinte.167.19.2103.

Eanes, R. 2015. *The newbie's guide to positive parenting* (2nd Ed.).

Epstein, R. M. 1999. "Mindful practice." *JAMA* 282 (9): 833–839.

Ermak, L. Jan. 26, 2014. "Beating the burnout: Nurses struggle with physical, mental and emotional exhaustion at work." *Holland Sentinel.* Retrieved from http://www.hollandsentinel.com/article/20140126/NEWS/140129418

Fahrenkopf, A., Sectish, T., Barger, L., & Sharek, P. Feb. 28, 2008. "Rates of medication errors among depressed and burnt out residents: Prospective cohort study." *British Medical Journal, 33*. Retrieved from http://www.bmj.com/content/336/7642/488.

Farber, S. April 2004. *The Radical Leap: A Personal Lesson in Extreme Leadership*. Chicago, IL: Dearborn Trade Publishing.

Foucault, M. 2003. *The Birth of the Clinic: An Archaeology of Medical Perception*. (A. M. Sheridan, Trans.). New York, NY: Routledge Classics.

Freudenberger H. Jan. 1974. "Staff burn-out." *Journal of Social Issues* 30 (1). Retrieved from http://onlinelibrary.wiley.com/doi/10.1111/j.1540-4560.1974.tb00706.x/abstract.

Gee, J. P. 2009. *The Anti-Education Era: Creating Smarter Students through Digital Learning*. New York: Palgrave/Macmillan.

GE Healthcare. Sept. 27, 2011. "Care for the caregiver." *The Pulse on Health, Science, and Technology*. Retrieved from http://newsroom.gehealthcare.com/patient-safety-care-for-the-caregiver.

George, B. & Sims, P. 2007. *True North: Discover Your Authentic Leadership*. San Francisco, CA: Jossey-Bass.

Gerteis, M., Edgman-Levitan, S., Daley, J., & Delbanco, T. L. (Eds). 1993. *Through the Patient's Eyes: Understanding and Promoting Patient-Centered Care*. San Francisco, CA: Jossey-Bass.

Gibson, R. & Prasad Singh, J. 2003. *Wall of Silence*. Washington, DC: Lifeline Press.

Givens, J., Tjia, J. Sept. 2002. "Depressed medical students' use of mental health services and barriers to use." *Academic Medicine* 77 (9). Retrieved from https://www.ncbi.nlm.nih.gov/pubmed/12228091.

Goitein, L. Jan. 11, 2001. "Market forces and graduate medical education." [Presentation at Internal Medicine Resident Conference, Massachusetts General Hospital].

Gottschall, J. 2012. *The Storytelling Animal: How Stories Make Us Human*. New York: Harcourt Mifflin Harcourt.

Gottschall, J. 2013. "Infecting an audience: Why good stories spread." *Fast Company*. Retrieved from http://www.fastcocreate.com/3020046/infecting-an-audience-why-great-stories-spread.

Gould, B. E., O'Connell, M. T., Russell, M. T., Pipas, C. F., & McGurdy, F. A. 2004. "Teaching quality measurement and improvement, cost-effectiveness, and patient satisfaction in undergraduate medical education: The UME-21 experience." *Family Medicine 36*, S57–S62.

Grissinger, M. 2010. "Transforming the culture of patient safety in organizations." *Pharmacy and Therapeutics 35* (6): 308–316.

Groff, J., Howells, C., & Cranmer, S. 2012. "Console game-based pedagogy: A study of primary and secondary classroom learning through console videogames." *International Journal of Game-Based Learning 2* (2): 35–54.

Gross, J. J. & Levenson, R. W. 1995. "Emotion elicitation using film." *Cognition and Emotion 9* (1): 87–108.

Gunderson, A. 2008. "Masters of science in patient safety leadership: Responding to the demands for a safer health system." *Journal International Association of Medical Science Educators 18* (1).

Gunderson, A., Mayer, D., & Tekian, A. 2007. "Breaking the cycle of error: Patient safety training." *Medical Education 41*, 518–519.

Gunderson, A., Tekian, A., & Mayer, D. 2008. "Teaching interprofessional health science students medical error disclosure." *Medical Education 42*, 531–532.

Gunderson, A., Smith, K. M., Centomani, N., McDonald, T. & Mayer, D. B. 2009. "Teaching medical students the art of medical error full disclosure: Evaluation of a new curriculum." *Teaching and Learning in Medicine Journal*. Retrieved from https://www.ncbi.nlm.nih.gov/pubmed/20183343.

Hallbach, J. L. & Sullivan, L. L. June 2005. "Teaching medical students about medical errors and patient safety: Evaluation of a required curriculum" [Research report]. *Association of American Medical Colleges 80* (6): 600–606.

Halbach, J. & Piscke-Winn, K. 2012. "Communication using dominoes" [TTE Emerging Leaders in Patient Safety Teaching Strategy].

Hartwell, E. P. & Chen, J. C. 2012. *Archetypes in Branding: A Toolkit for Creatives and Strategists*. Blue Ash, OH: How Books.

Hendren, R. March 29, 2011. "Nurses too scared to speak up." *Health Leaders Media*. Retrieved from: http://www.healthleadersmedia.com/nurse-leaders/nurses-too-scared-speak.

Hojat, M. 2004. "An empirical study of decline in empathy in medical school." *Medical Education* 38, 934–941.

Hospital Safety Score. 2015. "Home." Retrieved from http://www. hospitalsafetygrade.org.

Houston, T. K., Allison, J. J., Sussman, M., & Horn, W. 2011. "Culturally appropriate storytelling to improve blood pressure." *Annals of Internal Medicine* 154 (2): 77–84.

Institute for Family-Centered Care n. d. "Patient-and-family-centered care." Retrieved from http://www.ipfcc.org/pdf/CoreConcepts.pdf.

Institute of Medicine. 1999. *To Err Is Human: Building a Safer Health System* (L. T. Kohn, J. M. Corrigan, & M. S. Donaldson, Eds.). Washington, DC: National Academies Press.

Institute of Medicine. 2001. *Crossing the Quality Chasm: A New Health System for the 21st Century.* Washington, DC: National Academy Press.

Institute of Medicine. 2003. *Health Professions Education: A Bridge to Quality.* Washington, DC: National Academy Press.

Institute for Healthcare Improvement. 2016. "The IHI triple aim initiative." Retrieved from http://www.ihi.org/engage/initiatives/tripleaim/ pages/default.aspx.

Interprofessional Education Collaborative Expert Panel. 2011. "Core Competencies for Interprofessional Collaborative Practice: Report of an Expert Panel." Washington, DC: Interprofessional Education Collaborative.

ISMP. June 2013. "Disrespectful behavior in health care … Have we made any progress in the last decade?" Retrieved from: https://www. ismp.org/newsletters/acutecare/showarticle.aspx?id=52.

James, J. T. 2013. "A new, evidence-based estimate of patient harms associated with hospital care." *Journal of Patient Safety* 9 (3): 122–8.

Joint Commission n. d. "Speak up campaign." Retrieved from: https:// www.jointcommission.org/speakup.aspx.

Josiah Macy Jr. Foundation. 2014. *Partnering with Patients, Families, and Communities: An Urgent Imperative for Health Care.* New York, NY: Josiah Macy Jr. Foundation. Retrieved from: http://macyfoundation.org/docs/macy_pubs/ JMF_ExecSummary_Final_Reference_web.pdf.

Kachalia, A., S. R. Kaufman, R. Boothman, S. Anderson, K. Welch, S. Saint, & M. A. Rogers. 2010. "Liability claims and costs before and after implementation of a medical error disclosure program." *Annals of Internal Medicine* 153 (4): 213–21.

Kalaichandran, A. "Suicide among physicians is a public health crisis." *The Huffington Post.* Retrieved from http://www.huffingtonpost.ca/ amitha-kalaichandran/physician-suicide_b_8665388.html.

Kaiser Permanente Garfield Innovation Center. 2012. "Innovation begins with an idea." Retrieved from https://xnet.kp.org/ innovationcenter/index.html

Kern, D. E., Thomas, P. A., Howard, D. M., & Bass, E. B. 1998. *Curriculum Development for Medical Education: A Six-Step Approach.* Johns Hopkins University Press.

Kim, S., Kaplowitz, S., Johnston, M. V. 2004. "The effects of physician empathy on patient satisfaction and compliance." *Evaluation and the Health Professions* 24 (3), 237–251.

Kingston, M. B. Feb 2016. Personal Communication with D. Ford.

Kohn, L. T., Corrigan, J. M., & Donaldson, M. S. 2000. *To Err Is Human: Building a Safer Health System.* Washington, DC: National Academies Press.

Koloroutis, M. 2004. *Relationship-Based Care: A Model for Transforming Practice.* Minneapolis, MN: Creative Healthcare Management, Inc.

Kreuter, M. W., Green, M., Cappella, J., & Slater, M. 2007. "Narrative communication in cancer prevention and control: A framework to guide research and application." *Annals of Behavioral Medicine* 33 (3), 221–235.

Kraman, S. S. & Hamm, G. 1999. "Risk management: Extreme honesty may be the best policy." *Annals of Internal Medicine* 131, 963–967.

Kron, F. W., Gjerde, C. L., Sen, A. & Fetters, M. D. 2010. "Medical student attitudes toward video games and related new media technologies in medical education." *BMC Medical Education* 10 (50).

Landro, L. Sept. 1, 2009. "Patients get power of fast response." *Wall Street Journal.* Retrieved from http://www.wsj.com/articles/SB10001424052970204047504574384591232 799668.

Larkey, L., Lopez, A. M., Minnal, A. & Gonzalez, J. 2009. "Storytelling for promoting colorectal cancer screening among underserved Latina women: A randomized pilot study." *Cancer Control* 16 (1): 79–87.

Latino Health Access, Visión y Compromiso, & Esperanza Community Housing Corporation. 2011. *The Promotor Model: A Model for Building Healthy Communities.* Los Angeles, CA: The California Endowment. Retrieved from http://www.visionycompromiso.org/wordpress/wp-content/uploads/TCE_Promotores-Framing-Paper.pdf.

Leape, L. L. Shore, M. F., Dienstag, J. L., Mayer, R. J., Edgman-Levitan, S., Meyer, G. S., & Healy, G. B. July 2012. "Perspective: A culture of respect, part 1: The nature and causes of disrespectful behavior by physicians." *Academic Medicine, 87,* 845–852. doi:10.1097/ACM.0b013e318258338d.

Leary, M. R., & Tate, E. B. 2007. "The multi-faceted nature of mindfulness." *Psychological Inquiry: An International Journal for the Advancement of Psychological Theory* 18 (4): 251–255.

Levy, P. Dec. 2011. "Don't wait for Washington" [Blog]. *Not Running a Hospital.* Retrieved from: http://runningahospital.blogspot.com/2011/12/dont-wait-for-washington.html.

Levy, P. April 2011. "Sully inspires and presses for action" [Blog]. *Not Running a Hospital.* Retrieved from: http://runningahospital.blogspot.com/2011_04_01_archive.html.

Levy, P. Jan. 30, 2012. "Comparability doesn't matter." *Not Running a Hospital.* Retrieved from http://runningahospital.blogspot.com/search?q=%E2%80%9CTransparency%27s+major+societal+and+strategic+imperative+is+to+provide+creative+tension+within+hospitals+so+that+they+hold+themselves+accountable.

Lindeman, S., Laara, E., Hakko, H., Lonnqvist, J. March 1996. "A systematic review on gender-specific suicide mortality in medical doctors." *The British Journal of Psychiatry* 168 (3). Retrieved from http://bjp.rcpsych.org/content/168/3/274.

Loeb, J. M. 2004. "The current state of performance measurement in health care." *International Journal for Quality in Health Care* 16, 5–9.

Ludwig, D. S., & Kabat-Zinn, J. 2008. "Mindfulness in medicine." *JAMA* 300 (11): 1350–1352.

Makary, M., & Daniel, M. 2016. "Medical error—the third leading cause of death in the US." *BMJ,* 353:2139.

Marx, D. n. d. Outcome Engenuity. Retrieved from: www.justculture.org.

Masters, G. & Forster, M. 1996. *Performance Assessment Resource Kit.* Melbourne, AU: ACER.

Mayer, D. Nov. 2012. "Educate the Young" [Blog]. Retrieved from https://educatetheyoung.wordpress.com/2012/11/09/ the-second-victim-in-health   care-why-care-for-the-caregive r-programs-are-needed/

Mayer, D., McDonald, T. Doyle, S. & Granzyk Wetzel. T. 2011. "TTE interdisciplinary patient safety roundtable: Medical students' daily narrative reflections." *Patient Safety and Quality Healthcare.* Retrieved from http://www.psqh.com/analysis/tellurid e-interdisciplinary-patient-safety-roundtable/

McDonald, T. B., Helmchen, L. A., Smith, K. M., Centomani, N., Gunderson, A., Mayer, D., & Chamberlin, W. H. 2010. "Responding to patient safety incidents: The seven pillars.'" *Quality and Safety in Health Care* 19 (6).

McQueen, A. 2011. "Understanding narrative effects: The impact of breast cancer survivor stories on message processing, attitudes, and beliefs among African American women." *Health Psychology* 30 (6): 674–682.

McQueen, A., Kreuter, M. W., Kalesan, B., & Kassandra, I. A. 2011. "Understanding narrative effects: The impact of breast cancer survivor stories on message processing, attitudes, and beliefs among African American women." *Health Psychology* 30 (6): 674–682.

MedStar Health. April 2015. "Please see me" [YouTube video]. Retrieved from https://www.youtube.com/watch?v=380MiMDoddI.

Mello, M. M., Boothman, R. C., McDonald, T., Driver, J. Lembitz, A., Bouwmeester, D., Dunlap, B., & Gallagher, T. 2014. "Communication and resolution programs: The challenges and lessons learned from six early adopters." *Health Affairs,* 331, 20–9.

Merriam-Webster. n. d. "Culture." *Merriam-Webster.com*. Retrieved from http://www.merriam-webster.com/dictionary/culture.

Modern Healthcare. 2015. "Modern health care." Retrieved from http://www.modernhealth care.com.

Nance, J. 2009. *Why Hospitals Should Fly*. Bozeman, MT: Second River Healthcare Press.

National Center for Cultural Competence. 2004. *Bridging the Cultural Divide in Health Care Settings: The Essential Role of Cultural Broker Programs*. Washington, DC: National Center for Cultural Competence. Retrieved from http://nccc.georgetown.edu/documents/Cultural_Broker_Guide_English.pdf.

National Patient Safety Foundation. 2015. "Free from harm: Accelerating patient safety improvement fifteen years after 'To Err is Human.'" Boston, MA: National Patient Safety Foundation Retrieved from http://c.ymcdn.com/sites/www.npsf.org/resource/resmgr/PDF/Free_from_Harm.pdf.

National Quality Forum. 2016. "Mission and vision." Retrieved from http://www.qualityforum.org/About_NQF/Mission_and_Vision.aspx.

Obama, B. 2009. "Remarks by the president to a joint session of congress on health care." The White House. Retrieved from https://www.whitehouse.gov/the-press-office/remarks-president-a-joint-session-congress-health care.

Paul, A. M. March 17, 2012. "Your brain on fiction." *The New York Times*. Retrieved from http://www.nytimes.com/2012/03/18/opinion/sunday/the-neuroscience-of-your-brain-on-fiction.html?pagewanted=all

Peabody, FW. 1927. "The care of the patient." *JAMA*. Retrieved from: http://cell2soul.typepad.com/files/the_care_of_the_patient-1.pdf.

Pennebaker, J. W. 2000. "Telling stories: The health benefits of narrative." *Literature and Medicine* 19 (1): 3–18.

Piper, L. E., 2011. "The ethical leadership challenge: Creating a culture of patient-and-family-centered care in the hospital setting." *The Health Care Manager* 3 (2): 125–133.

Planetree. 2009. "Patient-Centered Care Improvement Guide." Retrieved from http://planetree.org/wp-content/uploads/2012/01/Patient-Centered-Care-Improvement-Guide-10-28-09-Final.pdf

Plews-Ogan, M., May, N., Owens, J., Ardelt, M., Shapiro, J., & Bell, S. K. Feb. 2016. "Wisdom in medicine: What helps physicians after a medical error?" *Academic Medicine* 91 (2): 233–41.

Powell, S., & Stone, R. A. 2015. "New framework for creating a resilient organization using reflective practices." In G. Sherwood, & S. Horton-Deutsch (Eds.), *Reflective Practices for Transforming Organizations,* 215–241. Indianapolis, IN: Sigma Theta Tau International.

Raina, P., O'Donnell, M., Schwellnus, H., & Rosenbaum, P. Jan. 14, 2004. "Caregiving process and caregiver burden: Conceptual models to guide research and practice." *BMC Pediatrics* 4 (1). Retrieved from https://www.ncbi.nlm.nih.gov/pmc/articles/PMC331415/

Reason, J. Sept. 27, 2011. *The Pulse on Health, Science and Technology.*

Reid Ponte, P; Connor, M., DeMarco, R, & Price, J. July 2004. "Linking patient and family-centered care and patient safety: The next leap." *Nursing Economics* 22 (4): 211–215.

Reid Ponte, P., Peterson, K. 2008. "A patient-and-family-centered care model paves the way for a culture of quality and safety." *Critical Care Nursing Clinics of North America* 20, 451–464.

Ryan, R. M., Rigby, C. S., & Przybylski, A. K. 2007. "The motivational pull of video games: A self-determination theory approach."

Scott, S. D. 2009. "The natural history of recovery for the health care provider 'second victim' after adverse patient events." *Qual. Saf. Health Care* 18, 325–330.

Scott, S. D. Aug. 18, 2014. "Sometimes it hurts to care." *BJC Today Online.* Retrieved from http://www.bjctodayonline.org/Home/ForYourHealth/NewsArticle/TabId/133/ArtMID/540/ArticleID/223/Sometimes-it-hurts-to-care.aspx.

Selanders, L. n. d. "Florence Nightingale." *Encyclopedia Brittanica.* Retrieved from: https://www.britannica.com/biography/Florence-Nightingale.

Self, J. G. Feb. 19, 2016. "A decline in integrity in health care" [Blog]. Retrieved from: http://johngself.com/self-perspective/2016/02/a-decline-in-integrity-in-health care-2/

Sherwood, G. & Zomorodi, M. 2014. "A new mind-set for quality and safety: The QSEN competencies redefine nurses' roles in practice." *Journal of Nephrology Nursing* 41 (1): 15–24.

Shanafelt, T., Bradley, K., Wipf, J., & Back, A. March 5, 2002. "Burnout and self-reported patient care in an internal medicine residency program." *Annals of Internal Medicine* 135 (5). Retrieved from: http://annals.org/aim/issue/20002.

Shanafelt, T., Boone, S., Tan, L., Dyrbye, L., Sotile, W., Satele, D., West, C., Sloan, J., & Oreskovich, M. Oct. 8, 2012. "Burnout and satisfaction with work-life balance among US physicians relative to the general US population." *Archives of Internal Medicine* 172 (18). Retrieved from https://www.ncbi.nlm.nih.gov/pubmed/22911330.

Shelton, D. L. Oct. 7, 2015. "Family of a woman who died after a medical error joins hospital's safety panel." *The Chicago Tribune*. Retrieved from http://articles.chicagotribune.com/2011-10-07/health/ct-met-medical-errors-20111007_1_medical-errors-safety-panel-patient-advocates.

Sherwood, G. & Horton-Deutsch, S. 2015. "Transformational learning: Improving quality and safety through reflective pedagogies." In G. Sherwood & S. Horton-Deutsch (Eds.), *Reflective Organizations: On the Front Lines of QSEN and Reflective Practice Implementation*.

Smetzer, J. & Navarra, M. B. 2007. "Measuring change: A key component of building a culture of safety." *Nurse Economics* 25 (1): 49–51.

Stiegler, M. P. Jan. 29, 2015. "What I learned about the adverse events from captain Sully: It's not what you think." *The Cardiothoracic Surgery Network*. Retrieved from http://www.ctsnet.org/jans/what-i-learned-about-adverse-events-captain-sully-it's-not-what-you-think.

Sorra, J., Famolaro, T., & Dyer, N. 2012. Hospital Survey on Patient Safety Culture 2012 [user comparative database report].

Southcentral Foundation. 2016. Nuka System of Care. Retrieved from https://www.southcentralfoundation.com/nuka/

Steinkuehler, C. & Chmiel, M. 2006. "Fostering scientific habits of mind in the context of online play." In S. A. Barab, K. E. Hay, N. B. Songer, & D. T. Hickey (Eds.), *Proceedings of the International Conference of the Learning Sciences* (723–729). Mahwah NJ: Erlbuam.

Taylor, E. W. 2009. "Fostering transformative learning." In J. Mezirow, E. W. Taylor, and Associates. *Transformative Learning in Practice: Insights from Community, Workplace, and Higher Education.* San Francisco, CA: Jossey-Bass.

Thomas, N. Dec. 15, 2004. "Resident burnout." *JAMA* 292 (23). Retrieved from http://jamanetwork.com/journals/jama/fullarticle/199994.

Triolo, P. 2012. "Creating cultures of excellence: Transforming organizations." In G. Sherwood & J. Barnsteiner (Eds.), *Quality and Safety in Nursing: A Competency Approach to Improving Outcomes* (305–322). Hoboken, NJ: Wiley-Blackwell.

Toussaint, J. Oct. 2015. "Hospitals can't improve without better management systems." *Harvard Business Review.* Retrieved from: https://hbr.org/2015/10/hospitals-cant-improve-withou t-better-management-systems.

United States Department of Education. 2010. "The condition of education." *National Center for Education.*

Van den Bossche, P., Gijselaers, W., Segers, M., Woltjer, G., & Kirschner, P. 2011. "Team learning: Building shared mental models." *Instructional Science* 39 (3): 283–301. doi:10.1007/s11251-010-9128-3.

VitalSmarts, Aorn, & Aacn. n. d. "The silent treatment: Why safety tools and checklists aren't enough to save lives." Retrieved from: http://www.silenttreatmentstudy.com/

Wachter, R. July 2013. "In conversation with ..." J. Bryan Sexton, PhD, MA [Interview]. *PS Net (Patient Safety Network), AHRQ.*

Warren, N. July 2012. "Family advisors in the patient-and-family-centered care model." *MedSurg Nursing* 21 (4): 233 – 239.

Wenger, E., McDermott, R., & Snyder, W. M. 2002. *Cultivating communities of practice.* Boston, MA: Harvard Business School Press.

Woods, M. S. 2007. *Healing Words: The Power of Apology in Medicine.* Oakbrook Terrace, IL: Joint Commission on Accreditation of Healthcare Organizations.

Wu, A. n. d. "Caring for the caregiver." *Johns Hopkins Health Care Solutions.* Retrieved from https://www.johnshopkinssolutions.com/ solution/rise-peer-support-for-caregivers-in-distress/.

Wu, A. W., Boyle, D. J., Wallace, G. & Mazor, K. M. 2013. "Disclosure of adverse events in the United States and Canada: An update and a proposed framework for improvement." *Journal of Public Health Research* 2, 32.

Yukl, G. 2002. *Leadership in Organizations* (5th ed.). Delhi, India: Pearson Education.

Zak, P. 2013. "The future of storytelling" [YouTube video]. Retrieved from https://www.youtube.com/watch?v=q1a7tiA1Qzo.

ABOUT THE AUTHORS

Anne J. Gunderson, Ed.D, MS, GNP, is the Assistant Vice President of Education, Safety and Quality at MedStar Health. Dr. Gunderson holds a doctorate in educational leadership and is a board certified geriatric nurse practitioner. Dr. Gunderson is Associate Dean for Innovation in Clinical Education and Professor of Medicine at Georgetown University. Dr. Gunderson began her career in medical education at the Southern Illinois School of Medicine in the department of medical education. Three years later she was recruited to serve as inaugural faculty at Florida State University College of Medicine in the department of geriatrics. The medical school was the first new medical school created in over twenty years. The new start-up team was successful and it change medical education for the future. Dr. Gunderson returned to Illinois and accepted a faculty appointment in the department of medical education at the University of Illinois College Medicine. During her tenure at UIC she designed and created the Masters of Science in Patient Safety Leadership programs and served as the founding program director. In 2010, the University Of Cincinnati College Of Medicine recruited her to serve as the Associate Dean for Medical Education and Professor, Department of Medical Education. During her tenure at Cincinnati she created a new four-year curriculum with the help of over 150 faculty members. During this time, she also successfully went through a LCME accreditation. In 2013, Dr. Gunderson was awarded with an outstanding leadership in medical education commendation. Later that year, Dr. Gunderson was awarded with the AAMC medical education laureate award.

Dr. Gunderson has recently created and directs a new Executive Masters in Clinical Quality, Safety and Leadership program at Georgetown University. Dr. Gunderson has served as a leader for The Telluride

Experience since 2010. She has also served as an invited advisor for the AMA and the Lucian Leape Roundtable. She served 3 years as Chair of the Association of American Medical Colleges Central Group for Educational Affairs and the Group on Educational Affairs; and 4 years as the UGME section leader. Dr. Gunderson's research interests have led to PI/Co-PI roles on projects funded by the U.S. Department of Education, HRSA, D.W. Reynolds Foundation, John A. Hartford Foundation and multiple state level grants. Gunderson has authored over twenty manuscripts.

Tracy Granzyk, MS, MFA is a writer, filmmaker and healthcare content developer working to increase awareness around a culture of patient and provider safety and patient engagement throughout the healthcare continuum. She creates innovative healthcare related stories for web, print and film, incorporating knowledge and experience gained from a successful career across a wide range of healthcare delivery entities that include: health system administration, the biotech industry and medical research. The result of having worn multiple healthcare hats is a perspicacious and inclusive perspective on what makes healthcare tick and areas that still need stories to be told.

Tracy is currently serving as Founder and Director, Center for Healthcare Narrative at the MedStar Institute for Quality and Safety. She is adjunct faculty at the Georgetown University and full-time faculty at The Telluride Experience where her focus is to teach healthcare stakeholders, patients and families the value and the 'how to' in elevating and honoring the stories of patients and the healthcare professionals. In 2010, Tracy created a social media presence via the Telluride Experience Blog, Twitter and Instagram for attendees, colleagues and alumni of The Telluride Experience, giving all a place and a platform from which to share their stories; an effort that has globally expanded the reach of the curriculum and knowledge shared, culminating in the multi-authored and -audience nonfiction work, *Shattering the Wall*.

Tracy's passion for patient safety grew after meeting Helen Haskell and Patty Skolnik in 2007 while working as a writer on the Tears to Transparency film series. So impressed with the work these inspirational patient advocates were doing to protect others from similar harm, Tracy was inspired to do the same. In 2011, she left a successful sales and marketing career in the biotech industry to contribute her writing, content development, and marketing skills to protect patients and providers. Tracy has directed and produced a growing film portfolio, along with numerous writing projects. In addition to leading the Center for Healthcare Narratives, Tracy is currently at work on a creative nonfiction book, a short documentary on medical education, and a feature length documentary examining how health disparities affect the well-being of moms and babies.

David Mayer, MD, is Vice President of Quality and Safety for MedStar Health. He is responsible for overseeing the infrastructure for clinical quality and its operational efficiency for MedStar and each of its entities. Dr. Mayer also designs and directs system-wide activity for patient safety and risk-reduction programs. Over the course of ten years, he held numerous roles including co-executive director of the UIC Institute for Patient Safety Excellence, associate dean for education, and associate chief medical officer for Quality and Safety Graduate Medical Education.

Dr. Mayer also founded and has led The Telluride Experience for the last twelve years. Additionally, he co-produced the patient safety educational film series titled "The Faces of Medical Error: From Tears to Transparency," which won numerous awards including the prestigious Aegis Film Society Top Short Documentary Award. More recently, Dr. Mayer was presented with the 2013 Founders' Award from the American College of Medical Quality and the 2016 Patient Safety Movement Humanitarian Award. Based on his commitment to teaching, service, and patient advocacy, he also has received the University of Illinois/American Association of Medical Colleges Humanism in Medicine Award and was recognized by the Institute of Medicine in Chicago in 2010 with the Sprague Patient Safety Award.

He regularly presents and writes on topics related to quality and patient safety, and he has received grant funding from the United States Department of Education, the Agency for Healthcare Research and Quality, and the Anesthesia Patient Safety Foundation. Most recently, Dr. Mayer was named by Becker's Hospital Review as one of Fifty Experts Leading the Patient Safety Field.

This program has only been possible because of
the generous support and enthusiasm of:
The Doctors Company Foundation, CIR, COPIC, MedStar Health